D0934713

GEORGE GRENVILLE

GEORGE GRENVILLE

A Political Life

PHILIP LAWSON

CLARENDON PRESS · OXFORD
1984

Oxford University Press, Walton Street, Oxford OX2 6DP
London Glasgow New York Toronto
Delhi Bombay Calcutta Madras Karachi
Kuala Lumpur Singapore Hong Kong Tokyo
Nairobi Dar es Salaam Cape Town
Melbourne Auckland
and associated companies in
Beirut Berlin Ibadan Mexico City Nicosia

Oxford is a trade mark of Oxford University Press

Published in the United States
by Oxford University Press, New York

Library of Congress Cataloging in Publication Data
Lawson, Philip.
George Grenville, a political life.
Bibliography: p.
Includes index.
1. Grenville, George, 1712–1770. 2. Great Britain – Politics and government – 1727–1760. 3. Great Britain – Politics and government – 1760–1789. 4. Statesmen – Great Britain – Biography. 5. Prime ministers – Great Britain – Biography. I. Title.
DA501.G73L38 1984 941.07'092'4 [B] 83-21998
ISBN 0-19-822755-8

Typeset by Hope Services (Abingdon)
Printed in Great Britain
at the University Press, Oxford

Preface

The 1760s have been a field of rich pickings for historians since Sir Lewis Namier destroyed the Whig interpretation of history some fifty years ago. The abundant source material has been studied exhaustively to construct a picture of each ministry complete with its daily trials and tribulations. George Grenville's period as first lord of the treasury from April 1763 to July 1765 is no exception. The importance to the American revolution of the American Duties Act of 1764 and Stamp Act of 1765 has been repeatedly stressed and received as much attention as any other parliamentary legislation in this period. The important constitutional issues arising from the parliamentary battles over general warrants have also been examined in depth by several authors. Even Grenville's personal dealings with the king while in office have been studied and evaluated as a vital factor in the political instability of the 1760s. What seems to be lacking in the historiography of the period is an assessment of Grenville's career, before he became first lord in 1763, and some examination of his period in opposition during the last five years of his life, 1765–70.

Far from being a novice in politics at Westminster, pushed to the forefront of national life, Grenville had, by 1763, a long career in opposition and senior ministerial posts behind him. His performances in the Commons had established his reputation as a feared and respected debater well before 1760. The problem with writing a history of his early career is the lack of manuscript sources after the great fire at Wotton in 1820. The image that remains from the personal papers of the years 1712–60 is of a man whose whole life revolves around sittings of the House of Commons. Nothing of the trivia in his life at Wotton or in his London home remains. In one respect this is not such a great loss, for Grenville's life was devoted to politics, everything else came secondary. Thomas Pitt, his ardent follower, immortalized Grenville's daily life when he wrote:

He was a man born to public business, which was his luxury and amusement. An Act of Parliament was in itself entertaining to him, as was proved when he stole a turnpike bill out of somebody's pocket at a concert and read it in a corner in despite of all the efforts of the finest singers to attract his attention.

Such rigid personal habits would by necessity render any account of Grenville's early life a political narrative.

For the period in office and the opposition during the 1760s the manuscript position is better, though life at Westminster again dominates his papers and correspondence. These have given rise to many well-documented studies of the years 1763–65 when Grenville was first lord. Yet the intriguing aspect of these manuscripts covering his later life is that they show Grenville did not enter a political limbo after his dismissal in July 1765. On the contrary he was to play an active role in opposition politics until his death. He gathered a strong and loyal following around him, and, in addition to the Rockingham faction, made a significant contribution to the rise of party. His knowledge of Commons procedure was unrivalled, and in an age of interminable debates on practice and precedence it gave him an influence beyond that warranted by his following at Westminster. Grenville's position remained strong in the years 1765–70 for the simple reason that he was an effective leader in the Commons at a time when ministries were built around leading figures in the Lords. Indeed, his achievements in these years have been sadly neglected and underestimated.

Acknowledgements

A study of this nature by necessity relies heavily on work already done in the field, and to those historians on whose knowledge I have drawn I express my thanks. No less important to this book has been the financial help given by the Twenty Seven Foundation, and I am grateful for the generosity of its trustees. To those owners of private collections relevant to Grenville's life I owe a special debt, and would like to thank the following in particular: the Earl of Harrowby, Mrs D. C. Bruton, the Hon. Charles Strutt, the Marquis of Zetland, the Earl of Powis, Lt.-Col. Walter Luttrell, Mr F. H. M. FitzRoy Newdegate, Earl Fitzwilliam and his Trustees, the National Library of Scotland Trustees and the Trustees of the Huntington Library, California. To all those librarians and archivists up and down the country kind enough to help me with this work I also say thank you.

In conclusion I would like to pay special tribute to those involved in the nuts and bolts of this book. Professor Peter Thomas has been the guiding light behind all my work and this study is no exception. His advice and criticism of my views on Grenville here proved invaluable, and his work continues to be an inspiration to all those, like me, fortunate enough to work in this field of history. I would also like to express my gratitude to the two people charged with turning my scrawl into type, Siân Jones and Mary Wyman. Their patience was exemplary. And the last word of thanks goes to my wife to whom this book is dedicated, for everything.

Contents

Chapter I

From Wotton to Westminster 1713–1744

> Undoubtedly Mr Grenville was a first-rate figure in this country. With a masculine understanding; and a stout and resolute heart, he had an application undissipated and unwearied. He took public business, not as a duty which he was to fulfil, but as a pleasure he was to enjoy; and he seemed to have no delight out of this House, except in such things as some way related to the business that was to be done within it. If he was ambitious, I will say this for him, his ambition was of a noble and generous strain. It was to raise himself, not by the low, pimping politics of a court, but to win his way to power, through the laborious gradations of public service; and to secure to himself a well-earned rank in parliament, by a thorough knowledge of its constitution, and a perfect practice in all its business.

This was the glowing epitaph that Edmund Burke (MP for Wendover) bestowed on George Grenville in the House of Commons on 19 April 1774. Almost four years had passed since Grenville's death and Burke could not conceal his admiration for an old adversary, to whom, he felt, 'this country owes very great obligations'.[1] Yet until recently few historians could be said to be so generous in their assessments of Grenville's career in public service. Grenville was blamed by his contemporaries for inaugurating a policy towards the American colonies that resulted in their loss and little has been said in his favour since. Judged for posterity by his enemies, Grenville still appears in studies of the eighteenth century as 'a man of very weak understanding'; the scourge of the American colonists, concerned solely with asserting parliament's sovereign rights over its recalcitrant subjects across the Atlantic.[2] In the last twenty years,

[1] *The Works of the Right Honourable Edmund Burke*, ii. 37.
[2] Ilchester, *Henry Fox, First Lord Holland*, ii. 227.

however, a gradual revision of his role in eighteenth-century politics has taken place, shedding new light on a fascinating and enigmatic character.

In fact it is hard to believe how such a narrow view of Grenville's life in the era of the American revolution could have represented an accurate picture. For here was a politician whose career spanned thirty years at Westminster, and included tenure of the highest offices in the state. Certainly Grenville aroused extreme emotions in those with whom he came into contact. His rise to the post of first lord of the treasury is marked with declarations of devotion and detestation from those who crossed his path. To those who saw merit in Grenville's unstinting application to the business of his office and obsession with practice and procedure in the Commons, Grenville was a loyal friend and leader. For others, an overbearing personality and inclination to repeat arguments as though he were talking to children aroused such antipathy in those on the receiving end that they could hardly bear his presence. It was this less attractive side of Grenville's character that prompted George III to say that he would rather see the devil in his closet than his former first minister.[3] Too much attention to his personal feuds and dealings with America has, however, obscured the real scope of Grenville's achievements as an MP from 1741 to 1770. A closer examination of the mythology surrounding Grenville's career reveals that he overcame the animosity of his enemies and even the hostility of the king to leave his mark on all aspects of political life in the mid-eighteenth century. The secret of this success was explained by the diarist Horace Walpole after Grenville's death in November 1770: 'Mr Grenville was confessedly, the ablest man of business in the House of Commons, and, though not popular, of great authority there from his spirit, knowledge, and gravity of character.'[4]

The Grenville family had held land around Wotton Underwood in the west of Buckinghamshire since the twelfth century, and by 1770 the estate extended to some 2,500

[3] Brooke, *King George III*, pp. 103–22 gives a thorough account of the antipathy between Grenville and the king.

[4] *Memoirs of George III*, iv. 125.

acres yielding perhaps £4,000 per year. Untitled squires they may have been but Grenville influence was felt throughout the county, at every level of society. During their long tenure of the Wotton estate, the Grenvilles became renowned as improving landowners. Throughout George's fifty-eight years of residence there the process of buying land, enclosing and draining infertile ground, and general building continued unabated: it became 'a sort of mania with the family'.[5] There was also a strong tradition of representation at Westminster that George Grenville took for granted as a young man. His father, Richard, sat for Wendover and Buckingham town until his death in 1727, and George's uncle, Sir Richard Temple, held the county seat of Buckingham before assuming the title Viscount Cobham. For all its power and prestige in Buckinghamshire, however, the family had yet to play a prominent rôle in national politics, and up to the mid-1730s there was no indication that this could be achieved and the pattern of events changed.

Born on 14 October 1712, the second of five sons and a daughter to Richard Grenville and Hester Temple, George Grenville was destined from birth for a legal career. The task of representing the family in parliament went to the eldest brother, Richard. George enjoyed a happy childhood, with devoted parents and a closely knit family providing all the amusement and stimulus he required. At Eton and Christ Church, Oxford, he proved a conscientious, hard-working scholar; he did not gamble or drink heavily and left no hint of scandal behind him. Grenville's career in the legal profession took a predictable course: Inner Temple 1729, Lincoln's Inn 1734, and called to the bar in 1735. His only worry in an otherwise smooth and successful entry into the legal profession concerned money. In the transition from student to advocate George depended on the financial support of elder brother, Richard, and other wealthy relatives. On his father's death in 1727 George inherited £3,000 to be administered by his brother. Relations between the two were healthy but it irked George considerably to be reminded

[5] *Buckinghamshire Victoria County History*, iv. 233. One of Grenville's first tasks as an MP was to pilot through the House an enclosure bill for land in Wotton Underwood, *Commons Journals*, xxiv. 195 (4 May 1742).

time and again that his spending should not exceed the limits of his inheritance.[6] He endured this period of dependence with good grace but in his later years economy and thrift became the *modus vivendi* of George's family life. These restraints coloured his attitude to personal finances and eventually those of the nation too.

The striking feature of Grenville's early political career is that he held no sway over his own destiny. His entry into the House of Commons at the general election of 1741 arose from an act of vengeance begun by his uncle, Viscount Cobham, eight years earlier. Grenville's mother, Hester Temple, was Cobham's favourite sister and after her husband's death in 1727, Cobham had more or less adopted the Grenville children at Wotton, to the extent of making the eldest son, Richard, his heir. Cobham's wealth and driving ambition carried him into influential positions in both local and national politics. A Whig of the old school, he gave unstinting support at Westminster to Sir Robert Walpole's ministry until 1733, when Cobham decided to oppose legislation proposing an excise on wine and tobacco. The main battles over this proposal took place in the Commons, where the government's majority on 5 April 1733 was reduced to sixteen votes, prompting Walpole's decision four days later to postpone the legislation indefinitely.[7] Cobham lent his support to this resistance and along with several other peers launched a campaign against the legislation in the Lords.[8] The success of their efforts brought swift retribution from Walpole. On 15 April, Lords Clinton and Chesterfield were dismissed from their offices and in the next month Montrose, Marchmont, Stair, Bolton, and Cobham followed suit. The blow for Cobham and Bolton was doubly felt because both lost the colonelcies to their regiments as well. It is difficult to sum up Cobham's anger at this, though Professor Plumb does give some idea of his rage when he says of the demotion: 'This astounded society. A colonel's regiment, purchased for hard cash, was regarded as his freehold. To be so deprived for a political offence looked like robbery.'[9]

[6] Wiggin, *Faction of Cousins*, pp. 78–9.

[7] Langford, *The Excise Crisis*, p. 78. Walpole announced his decision to drop the excise on 11 April to the Commons.

[8] H. M. C. Carlisle, p. 107.

[9] *Sir Robert Walpole*, ii. 281.

Cobham sought his revenge at Westminster. He threw himself into the opposition against Walpole, and vowed to do everything in his power to bring about his fall. Walpole had an early warning of Cobham's intentions, for soon after his dismissal he and Bolton introduced measures into both Houses simultaneously to prevent the Crown from dismissing officers on or below the rank of colonel, except by court martial or parliamentary Address.[10] However, Cobham had no intention of shouldering the burden alone. Over the next eight years he used his immense wealth to create a following of his own in parliament. This group became known as 'Cobham's cubs' or the 'boy patriots'. Its base was family connection. Cobham's relatives included three families with strong political traditions: the Pitts, the Lytteltons, and the Grenvilles of Wotton. He intended that all their male offspring should enter parliament and support his cause, and in this scheme of affairs the Grenvilles occupied a special place for his sister had borne five sons. In 1735 Cobham could depend on the support of four relatives: Richard Grenville, MP for Buckingham, Thomas Pitt, MP for Bridgwater, William Pitt, MP for Old Sarum, and George Lyttelton, MP for Bridgwater. Three, William Pitt, Lyttelton, and Grenville, all made their maiden speeches in support of a place bill on 22 April 1735 and from that moment on continued an unrelenting opposition to Walpole's policies until his fall in 1742.[11] What distinguished this group from the rest of the opposition was the violence of its speeches and the loyalty of its members to Cobham and one another.

After the general election of 1741 the boy patriots gained several new recruits, including two more of the five Grenville brothers, George and James. George came into parliament for the constituency of Buckingham town, and his return was a straightforward affair. Electoral interest in the seat lay between Cobham and the Dentons of Hillesden. Each nominated one candidate and since 1715 there had been no contests. Cobham selected George for this constituency while his elder brother, Richard, moved on to secure his return for the county. The election cost George no more than a modest

[10] Foord, *His Majesty's Opposition*, p. 184. These measures were defeated.
[11] Sedgwick, *Commons*, i. 84.

£58 12*s* 7*d*.[12] What expectations did the second Grenville son harbour about a career at Westminster? Had he been given the choice it seems fair to say that he would have preferred to stay in the legal profession. Cobham 'brought' George Grenville into the House in the true sense of the word.[13] He intended George to swell the ranks of the existing group under his control and follow the lead of his more experienced relatives. Grenville's aunt, Christian Lyttelton, certainly thought it a matter of regret that so promising a legal career should be jeopardized by entry into parliament as an avowed opposition member. 'A bold stroke', she told Charles Lyttelton, 'for so young a lawyer to begin with being against the Crown', and blamed his friends and relatives for 'putting him on so hazardous a point'.[14] The loss of a secure future in law might be the cause of some regret but election to the House did have its compensations. Grenville entered the House at a crucial time in the opposition campaign against Walpole, and with the others in Cobham's group, played a leading part in the events that led to his downfall. Indeed, the Commons proved a perfect platform on which to demonstrate Grenville's natural talent as a debater. Speaking in the Chamber came easily from his training as a lawyer, and, with an expertise in economic and financial affairs allied to an invaluable knowledge of Commons' practice and procedure, the future at Westminster held some promise.

Grenville's first months in parliament were full of drama and incident. Members of both Houses assembled on 1 December 1741 with the fate of Sir Robert Walpole's long administration in the balance. The seeds of the government's downfall had been sown some two years earlier when, in October 1739, Walpole agreed to enter a war against Spain for which he professed no enthusiasm. The ministry's confidence and performance were sapped with the continued poor showing in the campaigns against the Spaniards, and this gave heart to the opposition who attacked on all fronts. Ministers suffered their worst scare in February 1740 when a place bill moved by the opposition met defeat by the very slim

[12] Wiggin, *Faction of Cousins*, p. 66.
[13] This is the word Genville himself used, *Grenville Papers*, i. 423.
[14] Wyndham, *Eighteenth Century Chronicles*, i. 97.

margin of 16 votes.[15] Many of Walpole's own supporters considered this a bad omen so close to a dissolution and subsequent general election. Fortunately the government enjoyed a respite when opposition leaders overplayed their hand in February 1741 by moving Addresses in both Houses to the king asking him to dismiss Walpole from his counsels for ever. In the Commons, the majority of opposition Tory and independent members had no stomach for such a personal attack, nor the prospect of government under Walpole's enemies, Pulteney and Carteret, and thus voted with ministers in an overwhelming rejection of the motions.[16] This provided Walpole with the fillip needed to fight the elections, and results showed that his administration could not yet be written off. The opposition failed to win a majority at the polls which some of its leaders predicted.[17] Victory in this contest went to Walpole, though the difference in numbers between government and opposition parties had narrowed considerably from those in the old parliament.[18] What finally sapped Walpole's support in the new parliament, however, was the impasse reached in the war against Spain and her allies. Set-backs for British forces in Europe and the West Indies provided opposition leaders with a ready-made campaign against Walpole's conduct of the war. Furthermore, ineffectiveness of British command abroad was compounded by dissension and loss of confidence within the ministry over how to end hostilities and combat the growing antipathy to Walpole's government at home.

The end came slowly thanks to a brave rearguard action by Walpole himself. When parliament reassembled in December 1741 he performed masterfully in answering critics of the government's war policy. On 18 December Sir Robert thwarted opposition in the lobbies.[19] Yet Walpole's inability to whip in his supporters and the drift away of independent support soon began to take its toll.

[15] *Commons Journals*, xxiii. 438.

[16] Ibid., p. 648.

[17] Coxe, *Walpole*, iii. 567.

[18] This is the conclusion reached by a leading authority on this period, Owen, *Rise of the Pelhams*, p. 7.

[19] *Commons Journals*, xxiv. 32–3. A report of the day's proceedings can be found in Walpole, *Letters*, i. 146–7.

In a number of contested election returns before the Christmas recess government candidates suffered defeat, and, worse still, Dr George Lee, one of Walpole's leading opponents, was elected chairman of the Committee of Elections and Privileges, the very body that supervised trials of controverted elections. The Christmas holiday offered little respite for Walpole. In the new year decisions on contested election returns continued to go against the ministry, eating away at its majority in the Commons. In a final effort to dislodge Walpole, opposition leaders therefore decided to relaunch the campaign against the handling of the war with Spain. In these dramatic circumstances Grenville appeared on the national political stage for the first time. On 21 January 1742, in a House filled to record capacity, Pulteney moved that a select committee of twenty-one persons be appointed to examine all papers relevant to the recent conduct of the war. Walpole responded to this ill disguised personal attack in spirited fashion, defending the record of his long administration and taunting the opposition with accusations that smear tactics were the only weapons left with which to attack his policies. His reply sparked off a memorable debate which terminated with a division at 11 o'clock in the evening, and the slim majority of three votes for the government.[20] Grenville impressed onlookers with his vigorous support of Pulteney's motion. As the diarist Horace Walpole pointed out, 'There were several glorious speeches on both sides; Mr. Pulteney's two, W. Pitt's and Grenville's . . .'[21] The House soon recognized Grenville's abilities too. In May 1742 his fellow MPs elected Grenville on to a committee with six others to investigate public accounts, with the generous salary of £1,000 per annum.[22] His years in the legal profession living on a meagre allowance honed a talent for financial stringency and administrative efficiency that few of his contemporaries could match. It enhanced his reputation in a House of Commons perennially concerned with the growth of public expenditure and stood him in good stead during a long ministerial career. This particular scheme did not progress beyond a second reading in the Commons

[20] The figures were 253 to 250 votes, *Commons Journals*, xxiv. 53.
[21] *Letters*, i. 165.
[22] *Commons Journals*, xxiv. 258. Grenville actually came fourth in the poll.

because of government opposition. First ministers tried to reduce the salary stipulation to £500 a year, and when this failed simply put a negative on the motion for the scheme to pass. Grenville's ambition and financial expectations perished in the division by 136 votes to 66.[23]

Not all was gloom and despondency for the boy patriots in this session. Indeed, the primary objective of their long term of opposition and Grenville's own election to the House was achieved on 2 February 1742 when Walpole announced his resignation. Eight defeats over election petitions and the failure to command a workable majority in the Commons finally convinced the first lord, and a reluctant king, that a change of ministers was inevitable. George II viewed the prospect with distaste, for any new or reformed ministry would, by necessity, include Walpole's enemies on the opposition benches. He knew that only the most able and influential spokesmen from the opposition could command the stable majority in the Commons that had eluded Walpole since the Christmas of 1741. The king considered several alternatives: to persuade leading Tories and independents to join the ministry; a reconciliation with his son Frederick as a means of gaining the support of his twenty or so followers; and even the entry to office of the most powerful family-based factions in the Commons like Cobham's cubs or the friends and relatives of the Duke of Bedford. The first hope proved in vain partly because of the king's detestation of Tory attacks on his beloved Hanover but also from an intrinsic Tory antipathy to the idea of surrendering political independence in return for office. The second was put into effect, but Frederick's coterie lacked the ability to form a strong administration or dominate the Commons. The third alternative George II found least appealing because he did not want to be dictated to in the Closet by a faction like the Cobhams whose members and anti-Hanoverian opinions he disliked. As a compromise therefore the king approached those opponents of Walpole who, while commanding respect and authority, were not of a particular faction or group he found objectionable.

[23] Ibid., p. 268. This contradicts the account by Sedgwick, maintaining that the motion was thrown out by the Lords (*Commons*, i. 83).

In Pulteney and Carteret, George II found the ministers who would command a majority at Westminster and not assail him with unpalatable demands and favours in the Closet.[24]

In view of Walpole's retirement so early in the new parliament, Grenville could have been forgiven for doubting the wisdom of Cobham's political judgement. A promising legal career seemed a high price to pay for one Commons' speech before the fall. Prospects of office as compensation appeared remote too after the king selected Pulteney and Carteret to lead the new ministry. Once the common objective of Walpole's resignation had been achieved, divisions as wide as those separating Sir Robert from his enemies appeared in the opposition ranks. Cobham played his part in this collapse. He allowed his personal prejudices and antipathy to Carteret to govern the attitude of his followers towards the new ministry, deeply offending the king and making entry to office virtually impossible. In Cobham's view, the claims of the boy patriots to high office could only be advanced by eclipsing Carteret's influence at Court and reversing the pro-Hanoverian policy pursued under Walpole. The means to this end took the form of prolonged but vain attempts to drive a wedge between the heir apparent, Frederick, and his father, George II, in the hope of cutting off Carteret's main power base at Westminster and driving the king into Cobham's arms.[25] Subtlety was not one of Cobham's most obvious characteristics and the factiousness of this plan gained the group few allies. In the short term the new ministers cleverly outflanked Cobham by restoring the regiment lost under Walpole and then, in July 1742, promoting him to the rank of field marshal. Without the original grievance to parade, what reason could Cobham now have for holding out but opposition for its own sake?

The remaining weeks of the spring session of 1742 certainly gave Cobham cause to regret his headstrong behaviour. Urged on by Cobham and his cubs the opposition attacked on three fronts, but met only frustration and defeat. Most importance was attached to the Committee of Inquiry into the last ten years of Walpole's rule moved on 23 March.

[24] For a much fuller explanation of these events see Owen, *Rise of the Pelhams*, chap. iii.

[25] Wiggin, *Faction of Cousins*, p. 97.

This appeared the opposition's strongest point and yet, in the division, the inquiry carried by a mere seven votes.[26] Cobham may have wanted the Committee to produce evidence that justified his eight and a half years of opposition but an increasing number of MPs seemed prepared to forget the past and allow Walpole a peaceful retirement. The joy of the Cobhams at having William Pitt elected to the Committee of Inquiry on 26 March quickly evaporated when it began examining witnesses. Loyalty to Walpole made it impossible for the Committee to extract information from its chief witnesses. As a last resort on 15 May the Committee introduced a Bill of Indemnity, sparing witnesses from prosecution should they give incriminating evidence. This was thrown out in the Lords, however, and without a cutting edge the Committee perished and with it Cobham's hopes of undermining Carteret.[27] Grenville and his colleagues took part in two other unsuccessful skirmishes that session: in March a Place and Pensions Bill moved by the opposition was allowed to pass the Commons only to suffer a heavy defeat in the Lords; and on 31 March the Commons rejected a move to repeal the Septennial Act by a majority of twenty votes.[28] These set-backs were a severe disappointment after the heady days of January and Febuary, and to make matters worse Grenville ended the session struggling against a respiratory illness that eventually forced him to spend the whole summer in the south of France. Matters could only improve after these weeks of frustration, for the Cobhams had discovered the hard way that a vote against Walpole did not mean a vote in their favour.

The political fortunes of the group fluctuated considerably over the summer. Expectations for the next session of parliament went no further than active opposition in the Commons with a hope of making inroads into the government's majority. The opposition was well equipped for such a task, with talented speakers like William Pitt, George Lyttelton, and Hume Campbell within its ranks. The

[26] *Commons Journals*, xxiv. 146. The voting figures were 252 to 245.

[27] The ups and downs of the Committee can be followed in Walpole, *Letters*, i. 211–33.

[28] *Commons Journals*, xxiv. 457; *Egmont Diary*, iii. 264, and Owen, *Rise of the Pelhams*, p. 110.

government raised opposition hopes too by raising Pulteney to the peerage as the Earl of Bath, leaving the leadership in the Commons to Henry Pelham who at that point did not hold cabinet rank.[29] Disappointment came in the autumn, however, when William Murray, one of the opposition's ablest debaters and a close friend of Grenville, accepted the post of solicitor-general in the new administration. His entry into government provided a great boost to morale for ministers and went some way to robbing the opposition of its debating advantage in the Commons. Grenville remained ignorant of these developments throughout his period of recuperation in France. Letters sent to him from England contained only matters of general interest and firm instructions to remain abroad during the coming winter: 'let us prevail upon you to give till next spring to the establishment of your constitution', wrote William Pitt in October 1742, 'and be assured you are conferring the highest obligation upon your friends, and upon none of them more than myself.'[30] Even William Murray said nothing of his defection in a letter to Grenville in November: 'You know as much of foreign news as we do,' he observed blandly, 'we are told affairs in Germany go very well, and as to domestic, they wait the meeting of the Parliament.'[31] None of his family and friends expected Grenville to disobey these instructions, but by the autumn George had had enough of France and expressed an eagerness to return home for the opening of parliament on 16 November. This plan he abandoned after pressure from his colleagues, but a letter from his brother Richard in late November finally convinced Grenville that he could delay his return to Westminster no longer. In it he not only learned that his 'hero [Murray] is not the hero he was when you left him' but also of a sound beating for the opposition in a division on the Address by 259 to 150 votes.[32] He left Paris immediately, arriving in London in the first week of December 1742.

It says something for Grenville's self-confidence that he

[29] He was paymaster of the forces.

[30] *Grenville Papers*, i. 13.

[31] Ibid., p. 17.

[32] For the division see *Commons Journals*, xxiv. 336–7. His brother's letter appears in the *Grenville Papers*, i. 19.

never doubted his presence would galvanize the opposition effort in the Commons. Experience limited to one recorded speech and an abortive Committee on Public Accounts exercised little restraint on his behaviour: it was a personal and family characteristic to give all or nothing. Certainly, his friends and relatives set great store by his reappearance at Westminster: 'I hope to hear you are perfectly recovered', wrote his younger brother in October, 'that you may be present at the opening of parliament, not to lose the opportunity of distinguishing yourself as remarkably the next sessions, as you did the last.'[33] After the failure on the Address Grenville required no bait to tempt him into the House, but to dispel any indifference his brother also underlined the fact that the opposition now had a campaign on the European war that might bring success. The critical issue concerned British and Hanoverian forces in Flanders, and their rôle in the support of Maria Theresa, Queen of Hungary, against her enemies. On 6 December the secretary at war moved for the supply necessary for the army across the Channel, prompting a spirited attack from William Pitt.[34] He questioned the strategy of dividing the war effort in Europe, maintaining that direct subsidies to Maria Theresa would ensure an earlier defeat of her enemies. The plea fell on deaf ears: following a brilliant maiden speech refuting Pitt's strategic objections by Grenville's friend William Murray, the government carried its motion by 120 votes.[35] It is difficult to believe that this result stemmed from anything else but an inherent weakness in Pitt's view of the continental war. With or without Grenville this 'peevish resolution', as one opposition sympathizer put it, stood little chance of attracting support from the uncommitted members in the House.[36]

The last chance for the opposition to make something of this campaign came on 10 December when a very full House of Commons debated a grant for payment of the Hanoverian troops. Unlike the debate four days earlier, the opposition now found widespread support for its objections to paying

[33] S.T.G. Box 22 (25–6), Thomas Grenville to George Grenville, 16 Oct. 1742.
[34] For this debate see the account in *Parliamentary History*, xii. 906–40.
[35] Walpole, *Letters*, i. 311.
[36] *Egmont Diary*, iii. 268.

troops from the electorate with British funds. George Grenville, obviously intent on making amends for his previous absences, delivered an excellent speech that encapsulated the opposition case against the grant. First, there were the technical points arising from the assumption by ministers that Britain should be responsible for the troops without prior reference to parliament. Second, he raised the strategic point that the Hanoverians had not yet done any fighting, and, in his opinion, would do none until they were deployed with Maria Theresa's forces. He accused the government of misinforming the House as to their military position and significance, charging that they were simply an expensive show of force. Third and last, he spiced the attack with a flavour of anti-Hanoverianism aimed directly at the Tory ranks in the House, and in an unashamedly sensationalized conclusion to his speech, he declared:

> It is therefore not only certain that these troops, these boasted and important troops, have not yet been of any use; but probable that no use is intended for them, and that the sole view of those who have introduced them into our service, is to pay their court by enriching Hanover with the spoils of Great Britain.[37]

Grenville spoke last in the debate and his rousing attempt to exploit the anti-Hanoverian prejudice in the Commons may have swayed several waverers when the division came on. But it did not prove enough to convince the majority that ministerial policy favoured Hanover against the national interest: in the lobbies the opposition suffered another defeat by 260 votes to 193.[38]

No further opportunities for a full-scale opposition attack arose that session. In the Commons on 10 March 1743 the Cobhams supported a brief skirmish by other opposition groups over alleged irregularities in the disposal of government contracts. Defeat by a margin of twenty-nine votes showed a marked improvement in performance, but it

[37] The quotation and account of this debate are taken from *Parliamentary History*, xii. 940–1054.

[38] Ibid., p. 1054. The feelings Grenville attempted to exploit are summed up in a quotation by Chesterfield: 'If we have a mind effectually to prevent the Pretender from ever obtaining this crown, we should make him Elector of Hanover, for the people of England will never fetch another king from Hence' Walpole, *Letters*, i. 312.

proved an isolated incident.[39] Parliament was prorogued on 21 April without further discomfort for the government. The king and Carteret left in triumph for the continent soon after, and the Cobhams seemed doomed to an interminable period of factious opposition with which the group became increasingly identified. George, again suffering from respiratory trouble, left for a holiday in Scotland. It proved more than a vacation, however, for he was to make friendships with two of his hosts, James Oswald and Gilbert Elliot, that lasted a lifetime and served him well throughout his political career. He seems to have impressed the Edinburgh political élite in particular, who freely complimented him on his ability and promise after his return to England.[40] This was real solace, for in the political world the summer brought no upturn in the prospects of the opposition. Success for the king on the battlefield at Dettingen in June, and achievements for Carteret on the diplomatic front, culminating in the Treaty of Worms on 2 September, made Cobham's hope of undermining the ministry appear a distant dream. Indeed what hope Cobham had for office in the future no longer rested on the efforts of his group. It was Carteret himself who sowed the seeds of his own destruction during the remaining months of the year. His inability to appreciate the necessity of cultivating the favour of his colleagues or ministerial followers in the Commons resulted in a series of damaging internal squabbles over places and policy towards the continental powers. In consequence, by the late autumn the cabinet had split into two camps with Carteret and his small following on one side and Henry Pelham, his brother the Duke of Newcastle, and a majority of the cabinet on the other.[41] It is ironic, therefore, that when Cobham's fortunes were at their lowest ebb he suddenly found himself being invited to negotiations with the Pelhams with a view to discussing his terms for entry into office. Walpole, now the Earl of Orford, encouraged the idea of enlisting Cobham's help to outflank Carteret in a letter to his friend and protégé Henry Pelham on

[39] *Parliamentary History*, xiii. 1–44.

[40] *Oswald Memorials*, pp. 374–81 and p. 475.

[41] For background to this division of interest see Owen, *Rise of the Pelhams* chap. v.

25 August.[42] Smitten by a mood of distrust for everything Carteret said or did, the Pelhams accepted his advice, negotiating with Cobham intermittently until the opening of parliament in December. They found agreement an elusive goal. No problem existed over offices for the group: the talks foundered because Cobham would not serve with Carteret and, as Pelham well knew, the king would never support an administration without him. Each side refused to compromise, and when it became clear to Cobham that not only would Carteret stay but also that the Hanoverian troops, attacked by his followers the previous December, would remain in British pay he angrily resigned all his military posts.[43] With this action the door to office slammed shut.

For once the prospect of opposition did not appear so unwelcome for the boy patriots. Cobham now knew of the divisions within the administration and at least one controversial policy to be introduced at the opening of parliament. The campaign against retaining the Electoral troops was a strong-point for the opposition. A tide of revulsion against the Hanoverians swept through the country during the autumn, fuelled by a very active opposition press. At its root lay a growing conviction that British interests were being sacrificed to protect Hanover and its German allies. The inactivity of the army since Dettingen did nothing to dispel this impression. On the contrary the failure to follow up this victory with a sweep against the demoralized French only served to confirm the belief that concern for Hanover prevented decisive action in the war. The opposition suffered no shortage of examples to support its case, especially after the Earl of Stair resigned as commander-in-chief of the British and Hanoverian forces on 4 September. He and his supporters bombarded their friends in Britain with stories, some true, some imagined, of the king's Germanic preferences in military and diplomatic affairs. Their accounts held special appeal for the opposition because, behind it all, lay the malevolent hand of Carteret. The hopes and aspirations of the opposition leaders for the coming session found expression at a meeting in the Fountain Tavern on 10 November. They

[42] The letter was originally intended to wish him luck after his appointment as first lord of the treasury, Coxe, *Pelhams*, i. 91–3.

[43] Walpole, *Letters*, i. 396.

decided to launch a full-scale attack on the retention of the Hanoverian troops and composed a whip for circulation among absent friends. The influence of Grenville and the other Cobhams is evident in the message which made no bones about the importance of attending the debate:

your friends . . . most earnestly request you not to fail being in Town, the 21st in all events, because we apprehend advantage may be taken, in case of a thin attendance, to bring on the Hanover Troops; surprise being the only means left, of carrying them through the House, we beg you will be so good to communicate these fears and this request to all our friends in your County.[44]

Eleven Whigs and eleven Tories signed the whip, with the former group containing the names of George and James Grenville, George Lyttelton, and William Pitt.

These high expectations found only partial fulfilment. On 1 December, the opening day of the session, the opposition committed a tactical blunder by attempting to put a negative on the Address, a procedure that had been last used in the reign of James II. William Pitt led the attack with a violent speech against Carteret's ministry, including an unfortunate denigration of the military capabilities of the Hanoverian troops and the king's part in the battle of Dettingen. If it was intended to pave the way for the later debate on returning the Electoral troops, Pitt must have regretted his words, for it had the opposite effect. The majority of members disliked the timing and the tone of the speech and in the division the government won a handsome majority by 278 votes to 149.[45] In a House lacking Henry Pelham, seeking re-election after his promotion to first lord, the government certainly had cause to celebrate this result. It provided much-needed confidence five days later when Edmund Waller moved to address the king to discontinue the Hanover troops in the pay of Great Britain.[46] Pitt again led the opposition forces and, to a degree, repeated his error of 1 December by concentrating on the xenophobic rather than the strategic objections to

[44] Newdegate MSS 2550.

[45] *Commons Journals*, xxiv. 482–3. For an account of this debate see *Parliamentary History*, xiii. 135–230.

[46] The account of this debate and quotations that follow are taken from ibid., cols. 135–46 and 232–74.

retaining the Electoral force. This tactic was probably determined by two star witnesses for the opposition, William Strickland and Charles Ross, who served as officers under Stair and freely aired their prejudices against the Hanoverians during the debate. When Grenville spoke he too fell into this trap, but at least in his conclusion he did try to point out that the troops represented an irreconcilable divergence of interest: 'I will never rest', he told MPs, 'until the electoral are separated from the regal dominions.' This was more likely to attract waverers to the opposition cause than earlier outbursts, but it did not prove decisive. The House rejected the motion by a majority of fifty votes.[47] The best that could be said of this result was that it represented a better showing than that of 1 December, and when the Committee of Supply met to vote the necessary grant a further improvement might be achieved. It remained clear, however, that a unique opportunity to exploit what Horace Walpole termed the government's 'doubtful point' had been squandered.[48]

For the Cobhams matters went from bad to worse. After defeat on the Hanoverian troops, they found themselves in disagreement over the next stage of the opposition campaign. In view of the government's reduced majority on 6 December, Cobham saw another chance of reviving his scheme to undermine Carteret's position at Westminster and advocated all-out opposition to the continental war. Supported by the Grenvilles, Cobham could see no merit in simply sniping at the administration: they were, as one contemporary wrote 'always for attacking the whole measure of the minister, and treated the affair of the Hanoverian forces as a very inconsiderable part of the whole, and of little further service than exasperating the people without doors'.[49] Pitt and Lyttelton on the other hand considered a blanket opposition to the war of little benefit to the opposition or the nation as a whole. They wanted to adopt a dual approach, in which they were 'opposed as ever to the employment of Hanoverians and subsidies to Austria' while at the same time 'willing to support the motion that war should only be continued in

[47] *Commons Journals*, xxiv. 487.
[48] *Letters*, i. 396.
[49] Glover, *Memoirs*, p. 19.

conjunction with Holland'.[50] In the short term Cobham won this argument, and George Grenville replaced Pitt as the group's leading spokesman in the Commons. On 15 December, Grenville, seconded by Lyttelton, moved that:

an humble address be presented to his Majesty, most humbly to beseech his Majesty, that, in consideration of the exhausted and impoverished state of this kingdom, by the great and unequal expence it has hitherto been burthened with, he will not engage this nation any farther, by acting as we have hitherto done, or by concerting further measures, or entering into further engagements with other Powers without first entering into an alliance with the States General of the United Provinces, for the supporting and carrying on thereof upon stipulated proportions of force and expence, as was done in the late war.[51]

The motion smacked of compromise between Pitt and the Grenvilles with its acceptance of war only in conjunction with the Dutch. But the strong, even impertinent, wording identified it with Grenville. In fact it proved too strong for the House to consider seriously. After William Murray made the sensible point that no ministry should be expected to reveal the state of its military preparations in public, Pitt, Lyttelton, and Grenville retired to modify the most offensive phrases in the motion. It brought scant reward; in the division the government enjoyed an easy victory by 209 votes to 132.[52]

Failure to make any headway in this debate cut deep into the unity of the Cobhams over the Christmas recess. The eagerly awaited Committee of Supply on continuing forces in Flanders fell on 11 January 1744 and when the debate came on Pitt and Lyttelton sought no compromise with the Grenvilles. Cobham could not convert them to opposing the continental war outright. They shared his desire for Carteret's downfall but as a means to this end favoured a piecemeal approach in their attacks on ministerial policy. Pitt and Lyttelton attended the House on 11 January but walked out of committee before the division. This disagreement made it an important day for George Grenville, who now became Cobham's official spokesman. If he harboured any reservations about assuming Pitt's mantle he concealed them well on this day. What Grenville lacked in dramatic

[50] Ilchester, *Henry Fox, First Lord Holland*, i. 103.
[51] *Commons Journals*, xxiv. 492.
[52] Newdegate MSS CR 136/B2539/20.

rhetoric he made up for by strength of argument. One account of the day's proceedings captures his style well: 'Mr. Winnington, as Paymaster of the forces, officially proposed in the committee of supply, a grant for the maintenance of 21,358 British troops in Flanders. This motion was resisted by some of the violent members of Opposition, among whom Mr. George Grenville was the most conspicuous.'[53] Grenville did not shrink from his self-appointed task of taking on the ministry single-handed. He not only attacked the tactical thinking behind involving British troops in a continental campaign but also questioned the ministry's predictions about the strategic intentions of the other European powers. He hoped, in his words, to 'enter into all the views that can be proposed, and shew them rash'.[54] The solution to Britain's predicament again rested with a British withdrawal from the Continent. 'It would be more advantageous', he told members, 'to grant a subsidy to the empress queen, than to supply her with troops; because for £1,200,000 she could bring into the field fifty-four thousand foot and eighteen thousand horse; we could only maintain for the same sum, twenty-five thousand foot, and twelve thousand horse.'[55] Apart from the financial logic behind this argument, Grenville also believed that a stronger Austrian force would provide a serious bulwark against French expansionism and weaken Prussia's threat against her territories. Grenville could not have performed with more determination, and was rewarded in the debate with the support of several independent members. Yet there were speakers on the opposition benches, like Dodington (MP for Weymouth and Melcombe Regis) who offered only qualified backing to Grenville's case.[56] This naturally lessened the effect of Grenville's speech, and when Pitt and Lyttelton skulked off before the division came on the effort seemed entirely wasted. The question was carried for the government by 277 votes to 165, a demoralizing margin for Grenville who had given his all.[57]

[53] Coxe, *Pelhams*, i. 124.
[54] *Parliamentary History*, xiii. 394.
[55] Coxe, *Pelhams*, i. 124–5.
[56] *Parliamentary History*, xiii, 398–9.
[57] Ibid., col. 393, where there is also a full list of speakers in the debate.

After this set-back healing the rift between Pitt and the Grenvilles became the top priority amongst Cobham's friends. It proved more difficult than expected. Despite the obvious lack of success, George Grenville remained true to Cobham's wishes of seeking Carteret's downfall by opposition to every facet of ministerial policy. Pitt and Lyttelton, on the other hand, 'were beginning to tire of independent isolation', and stood fast by their decision to adopt a pragmatic approach in their attacks on the administration.[58] The merits of Grenville's strategy had suffered a severe test on 11 January and seven days later Pitt underlined its shortcomings. On 18 January came the debate on the estimate for the maintenance of the Hanoverian troops.[59] In a full House Pitt delivered a well-considered speech, singling Carteret out as the prime instigator behind a diplomatic policy that harmed the nation and served only the interests of the Electorate and the favourite's ambitions in the Closet. Such criticism struck a chord on all sides of the House, including several of the government's own supporters, and the reduction of the ministry's majority to thirty-five votes reflected the sympathy in the Commons with Pitt's views. To Grenville, who kept silent during the debate, this result represented a painful lesson in his political eduction. On the question of tactics he had been outmanoeuvred by an experienced campaigner who chose his ground and timing to perfection. As one leading historian of the period writes: 'Carteret's own high-handed treatment of his ministerial colleagues and his persistent disregard for the susceptibilities of the House of Commons had already rendered problematic his continuance in power; the division of 18 January 1744 first made his fall inevitable.'[60]

If it galled Grenville to have his thunder stolen in this manner he maintained a discreet silence. Doubtless it was some consolation that he and Pitt desired the same fate for Carteret whether they agreed on the strategy to this end or not. In fact the opposition made little headway after this debate, for attendance and interest in the remaining

[58] Ilchester, *Henry Fox, First Lord Holland*, i. 103.

[59] The account of this debate is taken from *Parliamentary History*, xiii. 388–462. There is no record of Grenville having spoken on this day.

[60] Owen, *Rise of the Pelhams*, p. 211.

weeks of the session never reached the heights attained in debates on the Hanoverian troops. The gravity of the news reaching London in early February of a threatened French invasion to support the Pretender also discouraged opposition to the ministry. Only the most zealous adherents to the opposition cause, among whom Grenville was numbered, sought to capitalize on the scare. He attacked the ministry on one particular point arising from its preparations for war, during a debate on 28 February to consider the suspension of Habeas Corpus. In peacetime this proposal would have raised a storm of protest but at this point of verbal hostility between Britain and France few eyebrows were raised. The invasion fear was at its most intense and several Jacobite suspects had already been arrested. Nevertheless Grenville launched a single-handed assault on the measure after its first reading. He objected in the first place to the assumption that the French would invade: 'France intends to frighten, not to invade us', he said. Grenville then warned members against granting the ministry powers of arrest without trial: 'Shall a suspicion of the ministry be an argument for a suspension of the palladium of our liberties for a day?'[61] The speech seemed inappropriate for the occasion. Faith in the tenets of English law guaranteeing individual freedom was admirable but when threatened with invasion, as French activity indicated, the state took an accepted course to bolster national security. The House shared the ministry's anxieties. When Lord Barrington proposed postponing action on the measure during the second reading on 1 March the motion met a firm rejection by 181 votes to 83.[62] Grenville found this pill doubly difficult to swallow as Pitt and Lyttelton again walked out during the debate in a show of contempt for the motion.[63]

Relations between the Grenvilles and their erstwhile friends Pitt and Lyttelton had now reached their lowest ebb, and for a time it seemed that loyalties within the group would be strained beyond endurance. Mounting divisions within the ministry, however, provided a timely impetus for another attempt at finding common ground on which

[61] *Parliamentary History*, xiii. 672.
[62] *Commons Journals*, xxiv. 594.
[63] Walpole, *Letters*, ii. 10.

to unite the group. Lyttelton shouldered the task of go-between and chief strategist, and the army extraordinaries, to be laid before the House on 13 March, became the target for a projected attack. It was a sensible choice, for questioning the accounts – in particular a payment of £40,000 to the Duke of Aremberg in 1742 – ensured that past disagreements over the conduct of the war would not be resurrected.[64] The debate on the £40,000 payment took place on 19 March when Lyttelton moved that 'the issuing and paying the Duke of Aremberg the sum of £40,000 to put the Austrian troops in motion, is a dangerous misapplication of the public money, and destructive to the rights of parliament'.[65] It has been said that this charge carried little weight, for although the extraordinaries, running to £500,000, represented a large increase on those of previous years, it was a novel matter for the opposition to claim the right to determine how this sum should be apportioned.[66] However, this is not how Grenville and his friends saw it at the time. They believed that fundamental constitutional issues were at stake and attached great importance to the debate. On 21 April Grenville explained their motives to his brother Thomas:

What we have dwelt upon more particularly was the article of £40,000 paid for putting the Austrian troops in motion for the year 1742 when *such great actions* were performed. This money was paid without the consent of parliament out of monies voted by parliament for other uses, and though parliament called for an account of all extraordinary services last year which was laid before them, yet this was never communicated. It was moved that this was a misapplication of public money and destructive to the rights of parliament.'[67]

During the debate MPs paid little heed to such lofty ideals. Their main concern was to dispatch the business as quickly as possible, and in the division the government secured a comfortable majority by 232 votes to 144.[68] Yet two aspects of note provided encouragement for the Cobhams in the face of this defeat. First, there was the minority vote of over 140; a high enough figure to suggest considerable sympathy

[64] The extraordinaries are printed in *Parliamentary History*, xiii. 676.
[65] Ibid., col. 677.
[66] Owen, *Rise of the Pelhams*, p. 219.
[67] S.T.G. Box 191 (3-6).
[68] *Parliamentary History*, xiii. 680.

with the case against the extraordinary payment to the duke, and, second, the tone of Grenville's speech. In his short contribution to the debate Grenville offered a thinly disguised hand of conciliation to Pelham and the rump of Walpole's followers still in the ministry, complimenting 'the last ministry at the expense of the present, whom he charged with still greater disregard for the rights of this House than their immediate predecessors'.[69] From this it is clear that Lyttelton had done his job well. These were the opinions that he and Pitt first voiced in January. The policy of blanket opposition from Grenville had come to an end. In contradiction to Cobham's wishes, detaching the Pelhams from Carteret became the primary aim of the group over the next seven months.

The last parliamentary battle waged by the Cobhams on this issue took place on 10 April 1744 when Lyttelton moved a vote of censure on the £40,000 payment to the Duke of Aremberg.[70] The exchanges covered much the same ground as the debate of 19 March, with Grenville speaking in support of Lyttelton and Pitt rather than leading the attack. In contrast to the debate of 19 March, however, little comfort could be gleaned from the division that followed. Though the minority increased to 144, the government secured a vote of 259, a remarkable figure so late in the session and one of the largest majorities since the Christmas recess.[71] The session ended on an equally depressing note for the opposition. On 15 April Grenville supported a move by the Cobhams to establish a Select Committee 'upon the affairs of the navy from the year 1742 to the present time, in which all Mr. Mathews' complaints, the conduct and shortcoming of ships and above all the economy of the navy were intended to have been strictly examined into'. The government resisted these changes and, in Grenville's words, 'carried it to have a committee of the Whole House where they know it is absolutely impossible to enquire at all'. Little wonder after this and other defeats that session he was moved to comment that 'the plan at

[69] Coxe, *Pelhams*, i. 136. Coxe has misdated this debate as 13 March and Sedgwick (*Commons*, i. 83) has copied the error.

[70] *Parliamentary History*, xiii 698.

[71] *Commons Journals*, xxiv. 651–2.

present seems to be whenever anything is discussed and attacked as blameable to make parliament give it sanction'.[72] This gloomy conclusion reflected the tone of Grenville's correspondence over the remaining weeks spent in London. Not only did he feel disappointment at the failure to remove Carteret but the rift between the boy patriots had left Grenville's pride battered and bruised. Given the opportunity to speak for Cobham in the Commons, Grenville soon discovered that Pitt's behaviour would prove crucial to advancing his own career. In practice he accepted this subordinate role with a stubborn resignation. Looking back on this part of his life some twenty years later, however, Grenville summed up his disenchantment with the position of Pitt's lieutenant:

During all this time I still continued giving my support to Mr. Pitt, notwithstanding the many public proofs I received of his indifference, coldness, and slight of every wish and opinion of mine, in the midst of the nearest intercourse, and of the strongest professions of friendship.[73]

In the final break with Pitt in 1761, it is clear the lessons of 1744 had never been forgotten.

During the summer recess Grenville confided in his younger brother Thomas that 'endeavouring to carry our new Turnpike Act into execution' provided a welcome relief from the battles at Westminster.[74] He purposely distanced himself from Cobham and Pitt by spending the summer in Ireland, allowing dust to settle on the division that marked the early months of the year. What news Grenville imparted to friends over these months was gleaned from the newspapers. He remained ignorant of the political scene in London; writing shortly before the opening of the new session his plans appeared vague and ill-defined.

The miserable situation of the public you see in every newspaper [he told his brother] and it seems to me very difficult to tell exactly the means of redressing it . . . however we shall try this winter whether the experience of the past may not add some weight to the reasons so often urged, and whether the public calamity, in which all are involved alike, may not induce some little consideration in those whom interest alone can persuade.[75]

[72] This quotation and the account of the debate on 15 April are taken from Grenville's letter to Thomas Grenville, 21 April 1744, S.T.G. Box 191 (3–6).

[73] *Grenville Papers*, i. 426.

[74] Ibid., p. 26.

[75] Ibid., p. 30 (22 Oct. 1744).

Ignorance is bliss. While Grenville indulged his passion for improving the Wotton estate and composing such meaningless pleasantries changes took place in London that had an immediate bearing on his political career. From a seemingly impregnable position in April, Carteret found by November that neither the king's confidence nor a parliamentary majority could protect him from his enemies within the ministry. What Grenville and his fellow patriots failed to achieve over the previous two years, the Pelhams, with Hardwicke and Harrington's help, executed in six months. There were many areas of conflict between Carteret and other members of the Cabinet, affecting both policy and patronage. In view of its overriding importance to national security, however, it is not surprising that Carteret's downfall should follow a dispute over the deteriorating military situation on the Continent; in particular the necessity of a formal treaty of alliance with the Dutch. Since the French declared war in the spring Britain and her allies, Holland and Austria, had suffered set-backs on every front. By the early autumn a full Dutch mobilization became imperative to offset French advances into the Low Countries and the reluctance of Austria to maintain a second front in Northern Italy. Yet Carteret would not press his allies. He believed that existing British and Dutch forces would thwart French ambitions and, to compound his crimes, for a time appeared more concerned with the Prussian threat to Hanover. To the exasperation of his colleagues he disdained to bother himself with constructing an overall strategy or finding the means of financing his generous, but ineffective, subsidy policy. Matters came to a head on 1 November when Newcastle presented the king with a memorandum indicating that unless Carteret were dismissed and the Dutch treaty concluded a mass resignation would follow.[76]

The king took this ultimatum badly and tried hard over the next three weeks to prevent the inevitable parting with Carteret, now Granville after the death of his mother in October. He first sought to persuade Harrington that if his colleagues resigned, he and Granville could govern alone.

[76] This episode is examined in more detail by Owen, *Rise of the Pelhams*, chap. vi.

When this ploy failed the king then enlisted the help of the Prince of Wales in a vain attempt at persuading the opposition lords Chesterfield, Cobham, and Gower into serving with Granville. The terms of acceptance were generous to a fault, including the dismissal of all Walpole's supporters; but as Chesterfield pointed out in his reply to the king no matter how generous the offer they could not serve with Lord Granville, or under him'.[77] At this point the Pelhams must have realized that victory was within their grasp. Even before Frederick's advance to the opposition lords met a rebuff, Henry Pelham had conferred with Chesterfield and on 21 November the two reached agreement on a future 'broad-bottom' administration. No places were settled but the dismissal of the followers of Bath and Granville and the adoption of a more systematic war policy became the basic conditions of the alliance.[78] Pitt, Lyttelton, and the Grenvilles eventually came round to accepting the deal arranged by Chesterfield but there had been a few uneasy moments on the way. In the summer when the first tentative negotiation between government and opposition took place Cobham expressed grave misgivings about any arrangements with Walpole's old supporters. He refused to compromise or relax his prejudice against the old adversary: as late as November 'Cobham's spleen and positiveness created more disturbance than anything else'.[79] In the end it was a timely piece of diplomacy by Chesterfield that broke Cobham's resistance. Shortly before his meeting with Pelham on 21 November, Chesterfield acquainted Cobham with the blunt fact that his own followers, the boy patriots, fully supported the negotiation and unless he ceased creating difficulties Granville's dismissal would not be achieved.[80] Cobham gave way, declaring in exasperation 'do what you will, provided you take care of my boys, the Grenvilles'.[81] This united front carried the day. On 23 November the king, outflanked and downhearted, intimated to Hardwicke that Granville had agreed to resign.

[77] *Marchmont Papers*, i. 88.

[78] Ibid., p. 89.

[79] *Parliamentary History*, xiii. 988.

[80] The deal between Chesterfield and the Grenvilles was settled in late September, *Marchmont Papers*, i. 52.

[81] Ibid., p. 80.

The scene for Grenville's entry into office was now set. As with his election to parliament in 1741, Grenville exercised little influence over the events leading up to his entry into the new administration. When he arrived in London in early November and learned of the changes afoot, he fell in with the plan to form a broad-bottom administration embracing all shades of opinion as the best means of uniting the country behind a vigorous prosecution of the war. The division of offices came as something of a disappointment, and the Cobhams had to content themselves with promises of preferment for their number rather than immediate rewards. Cobham himself resumed command of his regiment but took no political post, Lyttelton became a lord of the treasury, and George Grenville had the offer of a place on the admiralty board. Grenville was not entirely happy with these arrangements for they excluded Pitt, the most powerful voice in the group. Indeed at first he rejected the offer of admiralty lord for reasons he explained later:

As Mr. Pitt was not included in this arrangement, and as I knew to what a degree his mind was indisposed to Lord Cobham, I declined this offer, and earnestly begged Lord Cobham to excuse me, the only reasons for which were my friendship to Mr. Pitt, and my apprehension of family uneasiness.[82]

Whether or not this is the wholy story is open to doubt. Grenville's 'friendship' with Pitt was hardly idyllic and it seems likely that practical considerations of power and prestige in the Commons weighed heavily in his decision to reject the offer. Grenville had suffered Pitt's antagonism in the House before and clearly preferred to remain at his side than face an encounter on opposite sides of the Commons chamber. In the end his resistance came to nought. Relentless pressure from Cobham and then his mother, coupled with Pitt's indifference to this show of loyalty, convinced Grenville that acceptance would in fact be in his best interests.[83]

In early December 1744 Grenville thus began his long career in public service under the Duke of Bedford, newly appointed first lord of the admiralty. Over three years of opposition Grenville had witnessed dramatic changes in both

[82] *Grenville Papers*, i. 424.
[83] Ibid.

national politics and his own life. The fall of Walpole followed by the sudden eclipse of Granville were no less eventful than Grenville's own rise to prominence as an opposition speaker and the fluctuating political fortunes that resulted from bitter divisions within the Cobham group. Grenville entered office in the same manner as he had sought election to the House – as a novice. His place owed a great deal to Cobham's influence and Pitt's oratory in the Commons. Nevertheless Grenville's own efforts should not be overlooked. His strength in debate, reputation for financial management and tact when dealing with opposing interests like Pitt and Cobham suggested that he possessed the talents to make a success of ministerial office.

Chapter II

Political Apprenticeship, 1744–1754

Grenville's entry to office coincided with a period of stability in British politics that lasted some ten years. Pelham's choice of recruits gave the ministry, even without Pitt's official support, a considerable debating advantage over the opposition, later to regroup under the Prince of Wales. The presence of Grenville, George Lyttelton, Edmund Waller, and George Dodington on the front bench, promised a long period of calm and security for ministerial policy in the Commons. Indeed, as late as 1753 Walpole said of the sterility of parliamentary sessions under Pelham's guidance that 'it is more fashionable to go to church than to either House of Parliament'.[1] The only threat to this impressive union of talent came from within. Pelham reaped the benefit of admitting powerful groups like the Cobhams and Bedfords into his councils, but at the same time he opened a Pandora's box of jealousies and rivalries concerning power and influence in the new administration. Until his death in 1754, Pelham found the problems of intrigue and internal bickering as great as any of those posed by the opposition or the nation's enemies.

In the early years of his career Grenville played his part in these dissensions. Three factors governed Grenville's conduct in office during Pelham's lifetime and they were never entirely compatible with holding an official post. First and foremost, Grenville never forgot his debt to Pelham for providing much-needed financial security. A place on the admiralty board carried an annual salary of £1,000 and Grenville appreciated the independence this offered from his elder brother. It also regulated his behaviour in disputes with Pelham and his brother, the Duke of Newcastle. Second, Grenville harboured a keen desire to progress to the highest and most lucrative posts in government service. His loyalty

[1] *Letters*, iii. 142.

had been bought at a price but Grenville was not in office long before he realized that it could also be used as a weapon to further claims for promotion. Third, he retained a blind obligation to serve the interests of the group of 'patriots' he joined on entering the House. His friendship and family ties with William Pitt, in particular, led him into frequent skirmishes with his ministerial colleagues and even soured relations within the Cobhamite group. Worse still, devotion to Pitt's cause often conflicted with Grenville's own best interest, something that he realized only many years later when he lamented, 'thus I continued in the same office till Mr. Pelham's death in 1754, giving what support I was able to those who never gave any to me'.[2]

Grenville took up his duties at the admiralty on 27 December 1744, with some strongly worded advice from his cynical elder brother: 'Give yourself no trouble about the parish officers, only take care of the normal ones that even if the public be not benefited by your administration, your family at least may.'[3] In the event George needed no reminding of his duty to the family or his political allies. On 1 January 1744 Grenville told his new Scottish friend, James Oswald, of his excitement at the changing situation. The letter gives an intriguing insight into how Grenville weighed official duties against his political connections and loyalties:

> I came to town the day before yesterday from my election which was over to my great satisfaction last Friday. As soon as ever I took my seat at the admiralty board which was yesterday, I made it my business to enquire about what you mention in your letter and I find the form is, that the king signifies his pleasure to that board to make out a commission for the navy office, which has not yet been done but is expected every-day. What it is that delays it I cannot tell, unless it is that there is something to be done first with regard to the arrangement for the person whose place you are destined to supply. The moment I hear anything further of it, and it shall be my care to enquire, I will send you word, in the meantime I wish you was in town, that I might talk over this new world with you; notwithstanding the love of novelty, I retain the same opinions I had in the old world.[4]

In attempting to please everyone Grenville soon found his loyalties strained between supporting government policy and

[2] *Grenville Papers*, i. 427.
[3] Ibid., pp. 33–4.
[4] Oswald MSS, chest iv, c.

the wishes of his friends, though at first there were few difficulties, other than acclimatization, about his office at the admiralty. In common with all other members of the board, except Admiral George Anson, Grenville possessed no specialized knowledge of naval affairs. He relied entirely in the first weeks of office on his capacity for efficient administration and easy assimilation of facts to compensate for his lack of technical expertise. Other board members, like the Earl of Sandwich suffered similar disadvantages, and yet by diligent application to official business shortcomings in experience of naval affairs were largely overcome, making the Duke of Bedford's term as first lord 'administratively successful'.[5] All board members had ulterior motives of course for their diligence. In the case of Grenville and Sandwich, it was to assist serving relatives. By mid-February Grenville had found the time between official duties and attendance at Westminster to compose two letters to his brother Thomas, a captain in Hamilton's Atlantic Fleet. In this correspondence there is clear evidence that Grenville had already begun to use his influence in the family's favour. It tells of the admiralty's plans for Hamilton's squadron and arrangements made specifically for Thomas Grenville's ship and his subsequent postings.[6] This pattern of guardianship continued until the younger Grenville's untimely death in the spring of 1747 during Anson's victory over the French off Cape Finisterre.

Grenville's fortunes at Westminster did not run so smoothly, and in the most part his problems were self-inflicted. A cryptic note in his wife's diary explains the difficulty: 'The following year, 1745, though in office, I engaged with Mr. Pitt and Lord Temple in opposing the measures of government.'[7] At the root of this seemingly suicidal behaviour was Pitt's absence from the ministry. Grenville refused from the outset to lend his full support to the government until Pitt accepted a place in what became known as the 'broad-bottom' administration. In the circumstances surrounding the reformation of the ministry at the turn of the year this position appears barely tenable. It was no secret that the

[5] Baugh, *Naval Administration in the Age of Walpole*, p. 81.
[6] S.T.G. Box 191 (3–6).
[7] *Grenville Papers*, i. 424.

king disliked Pitt intensely for his attacks on Hanover and opposed his promotion to any office of state. Offers of advancement were made but, as Walpole noted, Pitt liked nothing better than a game of cat and mouse: 'he asked for Secretary at War, knowing it would be refused – and it was.'[8] Patience and a little tact would probably have overcome the king's prejudice in a matter of weeks, but this was not the way of the boy patriots. They wanted Pitt in on their terms whether it offended the king or not. Grenville would not contemplate a difference of opinion with Pitt in the House and sacrificed his votes for the ministry in a gesture of defiance. This commitment did nothing to enhance Grenville's reputation or standing amongst his ministerial colleagues, and brought about some erratic performances in the Commons.

The first division against the new administration on 16 January gave notice of Grenville's intent. It concerned a procedural point arising from a vote on the estimate for the ordnance for the ensuing year. Lord Strange and the Tory Sir John Philipps, like Grenville a recent recruit to office, led the opposition and though the government won the division comfortably, ministers had to suffer the embarrassment of having all but two of the new office-holders in the minority.[9] There is nothing in Grenville's correspondence at the time to suggest he felt uneasy in his position of rebel but he could not have escaped some discomfort during debates over the continental war that occupied so much parliamentary time from January to Easter. Despite all that had been said against Carteret's policies by both the Pelhams and many of the new adherents to the broad-bottom administration, the possibility of initiating radical changes in the conduct of the war remained minimal. Under Pelhamite influence the cabinet quickly resolved that the likeliest method of obtaining the necessary supplies and support for prosecuting the war would be to unite the country and parliament behind an all-out effort at beating the French. The dilemma behind this resolution centred on presenting the policy in parliament as something different

[8] Walpole, *Letters*, ii. 66.

[9] A full account of this debate and the one that follows on 23 January can be found in Owen, *Rise of the Pelhams*, pp. 251–2.

from Carteret's approach to the war. It did not prove easy. Opposition speakers attacked the ministry throughout February and March with stinging reminders that this policy was simply old wine in new bottles. In the end further criticism and divisive motions were only prevented by an unexpected declaration of support for government policy from William Pitt. On past record the mere hint of any increase in financial or military support for the continental powers would have aroused the whole Cobham clan into a state of frenzy. The group had attacked the use of Hanoverian troops, the diplomatic concessions to the Electorate, and latterly the subsidy policy to friendly European states when it became clear the Dutch were nervous of joining the fray. Yet when the Committee of Supply discussed increasing the deployment of troops in Flanders on 23 January, Pitt delivered the most forceful and cogent arguments for allowing the larger sum.[10] Opposition begun earlier in the debate evaporated after Pitt's speech. A similar pattern of events occurred the following month when the Committee of Supply met on 18 February to discuss an increased subsidy (of £200,000) to Maria Theresa. Ostensibly the larger grant was for payment of Hanoverian troops recently with British forces but now transferred to Austrian service.[11] Pitt accepted this point and did not oppose the grant, arguing that now the Hanoverians were no longer serving alongside British troops the original bone of contention had been removed. With this flimsy explanation the whole Cobham group, in and out of office, became committed to a position quite at odds with its previous standpoint on the issue. In executing this volte-face it appeared that Pitt's egotism had finally got the better of him. Could any further declarations of intent regarding parliamentary policy be taken seriously? There was, however, some method behind this madness, and the quiet acquiescence of his friends in this revision of attitudes provides the key. Grenville and the others well knew that Pitt's entry to office would depend on the king, and what better way to curry royal favour than supporting the policy nearest and dearest to the sovereign's heart.

[10] *Commons Journals*, xxiv. 714.
[11] For this debate see *Parliamentary History*, xiii. 755.

The chance of a reconciliation was enhanced further when Pitt also resigned his post in the household of the Prince of Wales. Accusations of hypocrisy obviously seemed an acceptable price to pay for Pitt's promotion.

Large majorities for the ministry on subsidies and troop reinforcements that occurred subsequently in the Committee of Supply bear witness to Pitt's influence in the House.[12] Yet with or without his support, Pelham remained in a strong position, for he had done a masterly job in destroying opposition to the broad bottom. His success in recruiting from all sides of the House, especially Tories like Gower, Sir John Hynde Cotton, and John Pitt, left the opposition without any natural leaders or, for a few weeks, any common cause on which to unite its attack. Tory prejudice against governments and the holding of office did not disappear, but under Pelham's tactful stewardship of the Commons it certainly became diluted. He began as he meant to go on: in January offering support for a bill ascertaining the qualifications of justices of the peace and in April, in deference to the demands of Tory MPs, he offered ministerial impartiality in the creation of six new commissions of the peace. So pronounced were the voting patterns of a body of Tory MPs by 1754 that the ministry felt able to draw up a list of Tories who were inclined to support government policy.[13] Furthermore individual Tory MPs entered and left office without the stigma of traitor that existed in Walpole's time. A significant number did remain in opposition, however, and before the Easter recess of 1745 had cobbled together an alliance with the supporters of Granville, Bath, and the Prince of Wales. There was no danger to the administration from this opposition front, but events in the Commons on 20 and 21 March during the granting of supply provided Pelham with a timely warning against complacency. In response to Pelham's request for a vote of credit of £500,000 for the ensuing year the opposition mustered 54 votes to the government's 185; and in a fuller House the following day the minority increased to 109 in a division on the Report.[14] This ended Pelham's honeymoon period and

[12] See, for example, *Commons Journals*, xxiv. 771 and 823.
[13] Owen, *Rise of the Pelhams*, pp. 260–2.
[14] *Commons Journals*, xiv. 823.

it was doubtless some comfort to know that at least internal opposition to his leadership appeared to have subsided. In the division of 21 March the Cobhams made their first real show of loyalty when Richard Grenville acted as a teller for the government side.

This gesture did not immediately gain George Grenville access to the inner councils of the administration. The memory of his opposition to ministers in the early weeks of the session had not dimmed sufficiently to warrant the trust of Pelham and his brother the Duke of Newcastle. In fact Grenville and Pelham found themselves on opposite sides of the fence after the Easter recess when the House considered the involved case of a miscarriage of the fleet in the action off Toulon in February 1744. The two admirals, Mathews and Lestock, in charge of the British force had missed an excellent opportunity in this action of destroying a Franco-Spanish fleet on its way to Brest. Instead of combining to beat the inferior enemy force, they allowed their personal feud to dictate preparations for the engagement, with the result that the ships under their command remained separated and the Franco-Spanish force escaped without serious loss.[15] There was an immediate public outcry at home and a court martial appointed to examine the dispute between the two admirals. This process took considerable time and while the advocates were still gathering evidence for the hearings the House of Commons intervened. On 26 February 1745, Charles Selwyn, seconded by Velters Cornwall, moved for a Committee of Inquiry. Ministers at first wished to oppose the inquiry as the case had still to be heard in the court. Yet feeling in the House and the country at large ran strongly in favour of parliamentary action, and in the end the government agreed to a wide-ranging investigation into the conduct of all persons involved in the dispute. In his role as admiralty spokesman, Grenville gave the grudging official approval for the inquiry:

he wished a negative might not be put upon the question, though he owned an enquiry here would be attended with great difficulties. The first care of the present Admiralty, had been to have this business

[15] This episode is explained in more detail by Richmond, *The Navy in the War of 1739–48*, ii. 1–57.

examined into, but their power extended no further than to laying it before a Court martial, which might be greatly embarrassed either in acquitting those condemned by the public voice, or finding them guilty merely in compliance with it. A decision of parliament either way would have more weight, and be thought more impartial.[16]

As it happened, Grenville's feeling that the inquiry would be attended with difficulty proved correct. The Committee ground away at the evidence for over a month but reached no firm conclusions, other than a recommendation that the Commons address the king to hold courts martial for Mathews and Lestock, along with several officers. The House was back where it started with the case. Grenville himself accepted the judgement, for it confirmed the fears he expressed on 26 February, and on 9 April he moved the resolutions to set the courts martial in progress. In a calm and reasoned speech he asked members to allow the legal proceedings to take their course and avoid apportioning blame to individuals when all the facts had not come out.[17] This sensible view aroused opposition the following day when the Tory Robert Vyner moved an amendment omitting Mathew's name from the resolutions. What lent weight to this manoeuvre was the fact that Pelham, Winnington, and Murray from the cabinet supported the amendment and later voted in the minority. It was to his credit that Grenville's standpoint reflected the majority view, and the amendment met rejection by 218 votes to 75.[18] The incident did not sour Grenville's relations with Pelham further because the issue cut across party lines. On the contrary it strengthened his hand in office and the House of Commons, repairing much of the damage inflicted by his opposition in the early weeks of the session. As he told his brother in a report of the proceedings the following month, 'I have a seat here [the admiralty], which I think it is now settled I am to have till the next sessions.'[19]

After Grenville retired to Wotton for the summer news of events at home and abroad harboured nothing but gloom. On the Continent the French defeated the allied army at

[16] *Parliamentary History*, xiii. 1204.
[17] Ibid., cols. 1278–84.
[18] Ilchester, *Henry Fox, First Lord Holland*, i. 114.
[19] *Grenville Papers*, i. 37.

Fontenoy in May, and in the following weeks occupied the whole of Flanders previously in British hands. The Austrians too suffered defeat in Germany, releasing French forces supporting Frederick for further action against Britain. Encouraged by these calamities Charles Stuart, the young Pretender, landed in Scotland on 25 July and raised his standard against the Hanoverian succession. By the end of September his Jacobite army had defeated the English under Cope outside Edinburgh and plans were afoot for the march south. For the remaining months of the year England was threatened by a simultaneous invasion from the north by the Pretender's army and in the south by France and its fleet. To make matters worse, the king spent most of the summer months in Hanover at odds with his leading ministers over the conduct of the war. Though Granville, the former Carteret, officially retired in the winter of 1744 the king had never dismissed him from his councils, and in foreign affairs Granville's pro-Hanoverian sentiments still held sway in the Closet. Matters of state during the king's absence were left in the hands of a regency council, and this quickly came to reflect the divergence of opinion between the king and his ministers. Of the council, Bath, Tweedale, and Stair saw little danger in the Jacobite uprising and refused to sanction the vigorous action to defend the realm urged by Pelham and Newcastle. Deadlock ensued, and in desperation the Pelhams pleaded with George II to curtail his visit. In late August the king reluctantly agreed to leave for London when it became clear to him matters at home had reached an impasse. His return, however, brought the Pelhams little joy. The king remained hostile to their opinions on the conflict at home and abroad, and on 17 September even made an abortive attempt through Lord Harrington to form a new administration. In its wake only a sense of duty amidst the rebellion prevailed upon the Pelham brothers to remain in office.

Against this backcloth of impending disaster, parliament opened for the new session on 17 October. The Address, concentrating solely on the rebellion, passed without trouble, but, as Walpole noted, 'I don't think, considering the crisis, that the House was very full.'[20] True, many members from

[20] *Letters*, ii. 144.

Scotland and the north of England could not or dared not attend parliament. But this pattern of low attendance, lasting until the new year, also reflected the lack of confident leadership at Westminster. As long as Granville retained the king's ear, the Pelhams could be leaders in name only, and it was at this low point in ministerial fortunes that Pitt decided to force his hand in the quest for office. In usual fashion he chose the least tactful manner of approach and insisted on dragging the whole Cobham group into conflict with the king and his ministers. On 23 October Pitt moved to address the king to recall all British forces remaining in Flanders; the implication being that the troops would be better employed defending the threat to their own shores than wasting time on a continental entanglement in the service of Hanover. The motion offered a direct challenge to the views of Granville and the king, and it soon became apparent that a good number of those MPs present shared Pitt's opinion. In the division the government only scraped home by 148 votes to 136.[21] This result indicated that apart from many uncommitted members voting against Pelham, several office-holders had followed the Cobhamite example and entered the minority lobby.[22] Pelham and his brother saw the meaning behind their action immediately, and decided that Pitt must be brought in without delay and the troublesome Cobhamites finally appeased.

This proved easier said than done. When Henry Pelham met Pitt on 25 October his terms of entry proved too high, especially a demand that two office-holders in the royal household be dismissed. Such appointments concerned the king and demanding a change would have represented a personal affront to the sovereign's prerogative. Another condition, that Britain reduce its forces aiding the Dutch in Europe and concentrate on a naval war against France and Spain, had not been aired before. Whether or not it came out on the spur of the moment is difficult to tell; considering Pitt's unpredictable behaviour in the past this could be the case. Nevertheless it was a strategy he pursued

[21] This description of the debate, and the division are taken from the Marquess of Hartington's report to the Duke of Devonshire 24 October 1745, cited in Owen, *Rise of the Pelhams*, p. 285.

[22] There is a list of the defectors in *Malmesbury Letters, 1745–1820*, i. 7–8.

with relentless determination over the remaining months of the year, and Grenville quickly fell in with the plan. Indeed it became a smoke-screen for a much wider campaign against the ministry's handling of the hostilities at home and abroad. Frustrated at the failures in Flanders and Scotland, the whole Cobham group wasted no opportunity to deliver stinging attacks on the inadequacy of ministerial policy in dealing with threats to the nation's survival. For this Pitt and Grenville, in particular, came in for some harsh criticism from their contemporaries. In his report of the debate on 21 November, for example, Henry Fox could not conceal his disgust for their insistance on dividing the House over a motion to augment naval forces:

> What makes it still more extraordinary is that the Duke of Bedford, Lord Gower, Lord Sandwich, Halifax and every Lord of the Admiralty but Grenville, are as much or more against the motion and the measure of a war merely naval . . . So that to carry their point, everybody must be made to submit to Lord Cobham's fireside.[23]

Yet loss of this division by 81 votes to 36 did not deter Pitt and Grenville from their chosen line of attack.[24] In December Hessian troops arrived in England to reinforce the Duke of Cumberland's army marching north to meet the rebels. To Pitt and Grenville it made no sense to use foreign troops at home when British forces still served abroad, and on 10 and 19 December they launched attacks against the deployment of the Hessians. In the latter debate Grenville spoke strongly in favour of using British troops to defend the realm and later became involved in an exchange with Henry Fox, who claimed they were splitting hairs in an hour of crisis. This argument proved more persuasive than those of Grenville and his friends, and it soon became clear that the House would be grateful for assistance no matter what its origin.[25] In the division on 10 December the government secured a majority of 138 to 44 votes and nine days later emphasized its strong position on this point by dividing 190 to 44 votes.[26] This result illustrated clearly that the

[23] Ilchester, *Henry Fox, First Lord Holland*, i. 120.
[24] *Commons Journals*, xxv. 21.
[25] Ilchester, *Henry Fox, First Lord Holland*, i. 121–2.
[26] *Commons Journals*, xxv. 22 and 25.

Cobhams had won few converts to their cause since the attacks opened in October. Indeed, Fox found a new angle by which to condemn Pitt and Grenville by pointing out that 'even Lord Cobham in the Cabinet was for the Hessians';[27] and Walpole commented for good measure after the debate that 'dismission of the Cobhamites is pretty certain'.[28]

In view of such reports it is difficult to put forward any defence of Grenville's conduct. Opposition to ministerial policy in October could be seen as a tactical move to force Pitt into office. But by late December, continued attacks on the war effort appeared factious and disloyal, and, more important, they hardly enhanced Pitt's chances of gaining a post in the administration. Evidence provided by commentators like Fox and Walpole, however, neglect the simple fact that none of the Cobham group supporting Pitt was warned about his conduct, let alone threatened with dismissal. Moreover, in their hostile reports of Pitt's behaviour in these months, they give the mistaken impression that the Cobham group criticized every aspect of the government's response to the rebellion. A great deal of parliamentary time in late October and early November was occupied by debates on a proposal for prominent noblemen to raise regiments to assist regular forces in the suppression of the rebellion. Initially this was to be done at their own expense, but after their formation the intention was to transfer the burden to the state. On 1 November members discussed the proposition in the Committee of Supply and it met violent opposition, not from Pitt and Grenville this time, but Fox and Winnington who held the posts of lord of the treasury and paymaster-general respectively.[29] In fact until its final acceptance on 4 November by 155 votes to 132,[30] the government suffered many unhappy moments as opponents to the proposal, both inside and outside the ministry, manoeuvred behind the scenes to prevent what they saw as jobbery at the public expense.[31] Pitt and Grenville shared

[27] Ilchester, *Henry Fox, First Lord Holland*, i. 122.
[28] *Letters*, ii. 164.
[29] Ilchester, *Henry Fox, First Lord Holland*, i. 117–18.
[30] *Commons Journals*, xxv. 14. It is indicative of the strong line the Cobhams took on this issue that Richard Grenville acted as a teller during the division.
[31] Walpole, *Letters*, ii. 147.

these reservations about an uncontrollable increase in ministerial patronage and expenditure, but supported the measure throughout on the grounds that the nation could not afford to spurn the assistance of its loyal subjects in such perilous times. This approach to the question also contradicts the notion that Pitt and Grenville viewed the troubles in Scotland and Europe simply in terms of a naval struggle, unlike some of the king's advisers. They never underestimated the Jacobite threat, merely demanding that the battles at home be fought by British troops. When Pitt first considered launching his campaign on the conduct of the war he stressed that this must be done in a disinterested fashion, 'without any hostility'.[32] This resolution had not always been honoured in the weeks following; but through the rhetoric and xenophobia Pitt, Grenville, and the other Cobhams repeatedly urged the House to wake up to the seriousness of the situation facing the country and take firmer action. These were the views that originally persuaded the Pelhams that Pitt must be brought into the administration; and despite the events of the autumn, they never wavered in their intention. Here was the trump card in the fight against Granville's influence in the Closet, and in early January 1746 negotiations began afresh to secure Pitt's entry to office.

On this occasion the chances of an agreement looked much healthier. The initial approach came from Pitt and, in a gesture of goodwill, he modified his terms of entry considerably. The strategic conditions relating to the war in Europe became secondary to the request for places in the administration for his associates. The Pelhams welcomed this mellower stand, and confided to Pitt that they would do their utmost to overcome the king's deep-seated prejudice to his promotion. Pitt accepted this declaration of intent in good faith and throughout January refrained from opposition at Westminster. It is probable that Pitt's change of heart was prompted by the receding threat of invasion. By the new year the Duke of Cumberland's army had succeeded in driving the Jacobite forces back to Scotland and the menace of the French fleet seemed to dissolve with

[32] *Marchmont Papers*, i. 146.

the worsening weather. As Grenville told his brother on 28 December 1745, 'I agree with you as to the French invasion, which I cannot believe is any longer intended to be carried into execution now the rebels are got northwards.'[33] Even Pitt would have found it difficult to carp at the military campaign when the nation's fortunes appeared to have turned a decisive corner. Whatever the motive behind the silence in January, his hopes of preferment were dashed by the king. He saw the improvement in Britain's strategic position as an opportunity to rid himself of Pelham and Newcastle once and for all, not a signal for promoting his personal enemies. In the second week of February, therefore, George II began a last attempt to reinstate Granville and Bath as his leading ministers, but it proved a disastrous experience for him. Granville and Bath were willing to lead but no one of sufficient standing in the Commons could be persuaded to follow. 'The wildness of the scheme', as Walpole wrote on 14 February, 'soon prevented others, who did not wish ill to Lord Granville, or well to the Pelhams, from giving in to it.'[34] The king's efforts at reforming the administration resulted in a co-ordinated tide of resignations. Grenville and all the admiralty lords but one handed in their seals, as did the treasury board and most of the cabinet and other leading office-holders. The city of London also expressed its disgust at the change by withdrawing the offer of a £3 million loan, with the cry 'No Pelham, no money'.[35] Bath and Granville persevered until mid-February, when the former told the king that the only course open was reinstatement of the broad bottom. In the next week all those who resigned took office again, and Pitt finally took his place in the lucrative office of joint vice-treasurer of Ireland. The king had been forced to swallow a great deal more than his pride in the hour of defeat.

As a group the Cobhams appeared well satisfied with this settlement. Pelham met their conditions of support handsomely: apart from Pitt's promotion, Lord Barrington, a recent recruit to the Cobhamites, received a place at the admiralty and James Grenville took a seat at the board of

[33] *Grenville Papers*, i. 48.
[34] *Letters*, ii. 175.
[35] Ilchester, *Henry Fox, First Lord Holland*, i. 125.

trade. The ministry from this point enjoyed unrivalled success. The suppression of the Jacobite forces continued apace, and the government gained unanimous backing for the continental war from its own followers. The acid test of loyalty for the new converts came on 11 April when the House was asked to approve a grant of £300,000 for the maintenance of 18,000 Hanoverian troops during the ensuing year.[36] The reputation of Pitt and Grenville had been built on attacking such payments, but, in this instance, their conversion to the Pelhamite cause was marked by a full confession of past errors by Pitt. Walpole described this about turn well:

> Last Friday was the debate on this subject, when we carried these troops by 255 against 122: Pitt, Lyttelton, three Grenvilles, and Lord Barrington, all voting soundly for them, though the eldest Grenville, two years ago, had declared in the House, that he would seal it with his blood that he never would give his vote for a Hanoverian.[37]

With a couple of hiccups Pitt's conversion lasted until Pelham's death in 1754, undoubtedly helping the ministry to establish a large, stable majority in the Commons.

Grenville himself appeared less enamoured of the new arrangements as the session wore on. He wanted reward for his loyalty to the Pelhams during February, and nothing happened in the first half of 1746 to alleviate his disappointment at remaining in the admiralty. Grenville became particularly angry in July when his hope of gaining a more lucrative and prestigous post at the treasury was thwarted by Pelham. He took this action for fear of promoting another of the Cobhams until the king had become used to Pitt's presence in office. Two places became available, and Grenville was passed over for Henry Bilson Legge and John Campbell. In his own explanation of this affair Grenville paid little heed to Pelham's difficulties with the king. Grenville considered himself senior to Legge for two reasons: 'Mr. Legge came into the Admiralty after me, who from the time of my appointment had done the business of that Board in Parliament, in which Mr. Legge had taken little or no share.'[38]

[36] *Commons Journals*, xxv. 123.
[37] *Letters*, ii. 185.
[38] This and the quotation that follows are taken from *Grenville Papers*, i. 425.

On both points Grenville had good grounds for complaint. His promotion to the Admiralty did preceed Legge's appointment, and Grenville's efficient performance as admiralty spokesman in the Commons, seen the previous year, for example, in the trial of Admirals Lestock and Mathews, seemed a fair claim to the treasury post. Pelham's view of the situation was all that mattered however, leaving Grenville with the impression that his application to business had earned him scant reward. Indeed his first reaction was to leave the ministry, commenting 'I therefore considered this preferment over my head as an affront, and determined to resign my office in consequence of it.' In the end Grenville was deflected from this course of action by an offer of the governorship of Barbados for his brother Henry, and assurances from Pelham that 'I should be next to go into the Treasury, and that neither of the gentlemen now preferred should go out of it before me.'[39] He did not, however, accept the situation with good grace, remaining bitter at the treatment he received from all quarters. Yet, at root, having pinned his colours so firmly to Pitt's mast, Grenville had only himself to blame when the expected promotion failed to materialize. The point, as Pelham stressed in his interview with Grenville, was that Pitt's promotion represented the limit to which the ministry could favour the Cobhams. The king and many of the ministry's own supporters were not happy with his inclusion in the administration and in the circumstances, Grenville could count himself fortunate to have received such a strong promise of preferment in the future.[40]

Grenville spent the remaining months of the year hard at work on admiralty business. His duties ranged from routine administrative chores to the planning of naval strategy to combat the Franco-Spanish threat. On one occasion, for example, Grenville wrote to James Oswald pleading sympathy for a man seeking to redeem the seamen's ticket of his dead brother. This problem of conditions of service concerned Grenville all his life and the letter shows he was far from being the heartless bureaucrat. After asking

[39] Ibid., pp. 425–6.

[40] See, for example, the note in the Earl of Marchmont's diary, *Marchmont Papers*, i. 223.

Oswald to expedite the case he concluded: 'I beg you will send your clerk or anybody else with him to the proper offices and give him what assistance you can.'[41] Most of Grenville's time in the autumn of 1746, however, was taken up with making preparations for an invasion of Canada the following spring. This was Bedford's idea, and Grenville expressed great enthusiasm for the scheme, doing a great deal of spadework on the strategic implications for Britain and her American colonies of launching the expedition.[42] As it happened, he undertook the task in vain. On 24 November 1746 Bedford told Grenville that their plans had come to nought. Troops and ships intended for Canada were to be used in a raid on the coast of France, leaving him to conclude 'that the intended expedition is entirely laid aside'.[43] This did not come as a great surprise to Grenville, whose other responsibilities that autumn included a search for means of reducing or funding the burgeoning naval debt. The problem had been inherited from previous administrations and exacerbated by the long war, creating shortages of supplies in men and ships on all fronts. On 19 November Grenville advised Bedford that it would be best to see Pelham personally and 'lay in our claim immediately.'[44] He hoped the government would borrow to reduce the sum and, indirectly, contribute to the cost of building the ships ordered for the Canadian expedition. Bedford agreed to this, even though he knew by then that the scheme for invading Canada was in jeopardy, and told Grenville to see Pelham himself. He received a cold reception. Pelham refused to borrow to reduce the debt, for he feared this would drive up interest rates on the regular loans. Instead he turned to the resources of the South Sea Company for funding, and, until the end of the war, this arrangement just about kept the admiralty's finances steady.[45] Pelham did not, however, save the Canadian expedition, which was abandoned in the new year when

[41] Oswald MSS chest iv, c.

[42] *Bedford Correspondence*, i. 156.

[43] *Grenville Papers*, i. 56.

[44] *Bedford Correspondence*, i. 190. Grenville erred in believing that the high debt and the feasability of mounting the Canadian expedition were linked.

[45] For further background to this incident see Wilkes, *A Whig in Power*, pp. 111–13.

moves for a renewed effort on the Continent began to be discussed with the ministry.

Grenville experienced greater success in his dealings with parliamentary business. As the opening of the new session approached ministerial confidence in its ability to secure a large majority for its policies was at its height. A new opposition under Bath and the Prince of Wales was threatened but as yet, they had made no formal declaration or designated their spokesman. Grenville echoed this optimism in his report of the Address on 18 November 1746. There was opposition on two points, he told the Duke of Bedford: Sir John Barnard objected to the ministry's apparent complacency with regard to the continental war and Sir John Hynde Cotton opposed the suspension of Habeas Corpus. No one took up the former argument, and only Mr Sydenham supported Cotton, which, 'upon a division', Grenville concluded, 'they divided but thirty-five'.[46] Grenville's satisfaction at this turn of events increased the following month when he arranged the return to parliament of his favourite brother, Thomas, at Bridport in Dorset. By an accident of timing the seat became vacant at the moment of the new opposition at Westminster and Thomas Grenville contested the election with Martin Madan, a supporter of the Prince of Wales. In what Walpole described as the Prince's 'opportunity of erecting his standard', the elder Grenville displayed considerable talent in beating off the challenge.[47] There was no dominant interest in the borough, though Grenville could exploit the limited influence exercised by both the admiralty and the Duke of Bedford in the constituency. A letter to the duke in November indicates Grenville realized the importance of leaving no stone unturned:

Your Grace has a very considerable estate in the neighbourhood which, as it gives the inhabitants of that town some dependence upon you, must give them the strongest desire of obliging you. Mr. John Way, attorney-at-law there and one of their returning officers is, I am told, a steward to your Grace. Would it be too much trouble for me to beg

[46] *Commons Journals*, xxv. 190. The voting figures were 134 to 35. Grenville's report can be found in his letter to Bedford of 19 November 1746, *Bedford Correspondence*, i. 188.

[47] *Letters*, ii. 256.

a line or two from you to him in favour of my brother, the captain, as a mark of your kind opinion and approbation of him?[48]

This canvass combined with a payment of three guineas for each voter produced the desired result, as Thomas Grenville defeated Madan by 122 to 116 votes. It proved a costly exercise but the prize of four Grenville brothers in the House more than justified the outlay.

Joy at this event was short-lived. On 3 May 1747 Thomas Grenville died of wounds received in the action with the French off Cape Finisterre. It came as a grievous blow to the whole family, especially George who knew that had his wishes been followed at the admiralty the younger brother would never have taken part in the engagement. In early March George had sought to obtain a separate command for his brother's ship, *Defiance*, and prepared the necessary orders to that effect. Bedford countermanded the order later that month, however, at the behest of Admiral George Anson, who was fearful of being left without sufficient firepower to engage the French fleet. A leading naval historian of the period says of the incident that Grenville then deliberately misrepresented an order for signature to extricate his brother from Anson's command.[49] The evidence for this assertion is taken from a letter sent by the Duchess of Bedford to Anson in early April.[50] In it she describes the sequence of events that led to Thomas Grenville's posting, and George Grenville's attempts to circumvent the duke's wishes. After Bedford had ordered Thomas Grenville to join Anson, she wrote, Grenville saw the duke in person to remonstrate with this order and make representations on his brother's behalf. The duke said nothing at the time but the following day sent Lord Vere Beauclerk, a fellow member of the admiralty board, to explain the reasons behind his decision. This Beauclerk did, though not to Grenville's

[48] Cited in Sedgwick, *Commons*, i. 232, in which there is a full account of this election.

[49] Baugh, *Naval Administration in the Age of Walpole*, p. 80. Baugh is mistaken when he says Barrington had a relative in the service (p. 81). It was Lord Sandwich, whose brother, the Hon. W. Montagu served as a captain at that time.

[50] This letter, sent in the first week of April, appears in Barrow, *Life of George, Lord Anson*, pp. 157–60.

satisfaction. The same evening, therefore, a third member of the board, Lord Barrington, went to the Duke of Bedford pleading Grenville's case. The duke remained unmoved and told Barrington that Grenville would have to accept the situation. At this point the story takes an interesting turn, for the duchess then says Barrington returned the following day with an order for the duke to sign that requested Anson not to detain Thomas Grenville's ship for more than seven days. 'This order', she told Anson, 'was treated with the contempt it deserved, and absolutely refused to be signed'; whereupon Barrington took another order from his pocket compatible with the duke's wishes which was signed immediately and sent off. The duchess concluded from this that 'the combination of Mr. Grenville and Lord Barrington are in to carry their favourite points by any means whatsoever'.

This account of the incident portrays Grenville in a shameful light. It is one that Anson's biographer, Sir John Barrow, has used to condemn Grenville, and it also provides the basis of Baugh's unflattering comments in his history of the admiralty during this period. Unfortunately neither have examined Grenville's account of the incident, where it can be seen that the root of the trouble was Bedford's insistence on using Vere Beauclerk as an intermediary. There are two essential differences between Grenville's account and that of the Duchess of Bedford. First, Grenville appears to have had real ground for complaint that Bedford countermanded the original order. As he told his brother on 2 April, 'the Admiralty has already promised me this cruise, with repeated vows and oaths'.[51] Second, on the evening Vere Beauclerk went to explain Bedford's reasons for overruling Grenville, he commented at some point that the *Defiance* could not be kept 'above a week' by Anson. As Beauclerk was Bedford's spokesman, Grenville intended to hold his superior to the seven-day limit, and that evening probably drafted an order including this stipulation. This represented insolent and arrogant behaviour but it was not a misrepresentation of the facts then known to Grenville. That Barrington took his

[51] *Grenville Papers*, i. 58–9. The account that follows and other quotations are taken from the same letter of 2 April.

orders to Bedford the following day is hard to believe. Once the duke had refused to sign the seven-day condition the matter would be quite straightforward, for Bedford had already issued the signed order instructing Thomas Grenville to join Anson. Though Grenville declined, other members of the board endorsed the order and saw to its dispatch. The sequel to this episode that suggests Grenville did endure a series of broken promises and groundless assurances came in the form of an apologetic letter from Anson to Grenville on 3 April. Anson appreciated the severe slight that Grenville suffered and asked him not to be among 'the number of my friends that neglect to write to me now and then when absent'.[52]

This gesture provided little comfort when Thomas Grenville's fate became known. After his posting to Anson's squadron George commented that 'I have not been at the board today, nor don't much care if I never go again': prophetic words in the circumstances, for in June, much to his relief, George Grenville's promotion to the treasury finally took place.[53] With the promotion came a welcome increase in salary to £1,400 per annum. A vacancy arose from the dismissal of Lord Middlesex who broke with the Court and threw in his lot with the new opposition under his friend the Prince of Wales. The necessity of declaring loyalties came to a head because Pelham decided to dissolve parliament and call a general election. In the fight for seats in the new House loyalties could no longer be divided between the king and the heir apparent. The official reason for the dissolution, Walpole pointed out, was 'the impossibility of making either peace or war, till they are secure of a new majority'. But he suspected, with some justification, that the true motive lay in an attempt 'to disappoint the prince, who was not ready with his elections'.[54] It proved a master-stroke. Despite an effort to cobble an agreement with the Tories, the prince's supporters suffered badly at the polls. In his analysis of the results, the Duke of Newcastle predicted a government majority of well over a hundred votes.[55] The electoral

[52] *Grenville Papers*, i. p. 61.
[53] Ibid., p. 59 and p. 426.
[54] *Letters*, ii. 277.
[55] Owen, *Rise of the Pelhams*, p. 317.

fortunes of the Cobhams followed this encouraging trend. In his report to Richard Grenville of 30 June, Pitt declared triumphantly that 'elections in general go better than the most warm expectation could promise'.[56] The only blur on the horizon concerned Thomas Pitt's decision to abandon the group and revert to his old task of election manager for the prince. In Buckingham the Grenvilles took control of the town, securing the return of both Richard and George. The agreement to share the nomination with the Denton family had lapsed in 1744 when George Grenville joined the ministry and George Denton remained in opposition. The strength of Grenville influence in the borough was such, however, that it precluded any serious opposition in this or any other election that century. In his report of the Buckingham election to Pitt, Richard Grenville underlined the uphill struggle Denton faced in the election. He offered £700 a vote but 'not one six pence offered, promised, or given by me and not a convert made against me', boasted Richard to Pitt, 'is it not an honour to represent such a corporation?' It is little wonder Denton declined the poll, and the elder Grenville did not exaggerate when he wrote after the election, 'I am now indeed Lord of Buckingham not for life only, but for me and my heirs with remainder to all my brothers and this issue.'[57]

Parliament reassembled for the new session on 10 November 1747. The Address passed without difficulty and until the Christmas recess MPs occupied themselves with election petitions. In a division on the Seaford election on 20 November, Newcastle's prophecies came to fruition, when the government beat the opposition in a well-attended House by 247 to 96 votes.[58] Ten days later in a debate on supplies for the ensuing year this performance was bettered with the government enjoying success by 239 to 54 votes.[59] Attendance at Westminster for the remainder of this and the spring session of 1748 fell away now that the superiority of

[56] *Grenville Papers*, i. 66.

[57] The quotations regarding Buckingham are taken from the account in Wiggin, *Faction of Cousins*, p. 118. For the past year Richard Grenville had been cultivating the borough gaining the support of the burgesses qualified to vote in a parliamentary election.

[58] *Commons Journals*, xxv. 429.

[59] Ibid., p. 454.

government support had become apparent. Ironically, the only threat to Pelham's position came from debates initiated by the Grenvilles over the Buckinghamshire assizes.[60] The root of this problem lay in a struggle for prestige and influence in county affairs between the Grenvilles and their rivals, which should never have reached the Commons. Prior to the general election of 1747, Willes, the lord chief justice, ordered that the summer asizes held since 1720 in Buckingham be moved permanently to Aylesbury. At the time the reasons for this change remained a mystery, but to the Grenvilles it smacked of blatant electioneering by Willes whose son, Edward, was a candidate for Aylesbury in the forthcoming election. Bringing the assizes to the town twice a year not only meant status for the family but also increased trade for the shops and inns. There was no predominant interest in the borough, and Edward Willes's eventual success at the polls certainly owed a great deal to his father's action. Worse still from the Grenville viewpoint, Willes and his fellow candidate, Lord Inchiquin, intended to support the opposition under the Prince of Wales at Westminster. After the election the Grenvilles wasted no time in seeking redress for what they considered an abuse of Willes's legal power.

In true Grenville fashion the means adopted caused maximum offence. After a vain attempt in July 1747 to persuade the chancellor Lord Hardwicke, to bring pressure on Willes to reverse the decision the family decided to seek parliamentary support for their cause. The means to this end became a planned petition to the House of Commons on behalf of the inhabitants of Buckinghamshire requesting the return of the summer assizes to the county town. This would be presented by the Grenvilles to the Commons along with a motion for the necessary legislation to put the petitioners' demands into effect. The petition did not go well. Apart from the town of Buckingham, which had most to gain, other areas of the county expressed little enthusiasm for the Grenville cause. This reticence was later reflected in the reaction of MPs to the petition, and Richard Grenville's

[60] The account that follows is based on the material contained in Grenville's notes, Willis MSS 114.

motion for legislation on 12 February 1748. In a division to allow the Grenvilles to introduce a bill returning the summer assizes to Buckingham the opposition reduced the ministerial majority to sixty votes.[61] The reasons for this lukewarm reception soon became apparent when the bill itself came before the House seven days later. Sir William Stanhope, who sat for the county, led an attack on the proposal with a speech full of bitter invective. He called the bill 'the errantest *job* that ever was brought into Parliament', and declared that MPs should not be asked to approve legislation born out of one family's malicious efforts to dominate the town of Buckingham.[62] Stanhope was ably supported by two followers of the Prince of Wales, Thomas Potter and Dr George Lee, who cleverly undermined the case for returning the assizes to Buckingham. They laid stress on three points: first, the lack of an adequate gaol in Buckingham to secure prisoners awaiting a hearing; second, the inaccessibility of Buckingham for people from other parts of the county, and third, that the granting of the assize to Buckingham in 1720 had only been a temporary expedient never endorsed by charter or law. These were strong arguments and in response Grenville, Pitt, and Henry Fox made a poor showing. Grenville laboured the historical and practical justification for holding the assize in the county town. But many of his precedents were refuted by Thomas Potter, who displayed an unrivalled knowledge of the county's history and development. The best Grenville could offer was a plea on behalf of the tradespeople who had suffered through no fault of their own, and a promise of a new gaol. As the debate wore on the attacks began to range beyond the subject at hand. Distaste for the way the Grenvilles had inflated their local squabbles into an affair of national importance eventually turned into criticism of Pelham's leadership. It was bad enough allowing the Grenvilles to force Pitt on to the king, Robert Nugent said, now they were dictating ministerial policy to their own convenience

[61] The figures were 184 to 124 votes, *Commons Journals*, xxv. 512.

[62] The account of the debate that follows is taken from three sources: (a) *Parliamentary History*, xiv. 204–44; (b) Walpole's letter to Sir Horace Mann, 11 March 1748, *Letters*, ii. 306–7; (c) Ilchester, *Henry Fox, First Lord Holland*, i. 150. Stanhope's quotation is taken from *Parliamentary History*.

as well. In reply Pelham avoided an answer to these personalized attacks and merely blamed the lord chief justice for causing all the trouble with his decision to change the assizes. It was an unsatisfactory way to put an end to the debate and the result of the division provided no compensation. On a motion for reading the bill a second time the government beat the opposition by 182 to 112 votes.[63]

The matter dragged on for another three weeks. On 2 March counter-petitions from the bill's opponents came before the House and were ordered to lie on the table until the second reading set for 7 March. Though no account of this debate is extant, it seems clear that the opposition put up a good fight. In a division on the committal the minority still numbered over a hundred votes.[64] On 15 March opponents of the bill concluded the struggle with a similarly good showing on the third reading and the motion to pass, going to defeat by 155 to 108 votes.[65] This ended an unpleasant month for Grenville and the other Cobhams. If it concerned them at all, the episode had left their popularity at its lowest ebb. Grenville may complain that he was 'quite ashamed Parliament has thought nothing worth their earnest consideration but a private or at most provincial squabble in which my eldest brother and myself are personally concerned', but to many MPs their success in forcing Pelham's hand over the assizes was simply the culmination of three years of unpardonable behaviour towards parliament and the king.[66] In this saga Grenville's open, and unpunished, opposition to Pelham while at the admiralty, and Pitt's forced entry to office and the change of opinion on the continental war all fuelled a deepening resentment at the behaviour of the group. It was fortunate for the ministry that Grenville and Pitt kept a low profile for the remaining months of the year.

In fact Grenville left London for Bath immediately after the Assize Bill passed the Commons to recuperate from another bout of respiratory trouble. Henry Pelham kept him

[63] *Parliamentary History*, xxv. 530.

[64] Almon, *Debates*, iii. 158–60 does offer a summary of the petitions. For the division, which was 150 to 103 votes, see *Commons Journals*, xxv. 538.

[65] Ibid., p. 570. Amendments were made to the bill in the Lords and sent back to the Commons for approval. This took place after a further division on 6 April 1748, and the figures were 131 to 78 votes, ibid., p. 620.

[66] Oswald, *Memorials*, p. 385 (17 March 1748).

informed of government business in his absence, and was especially delighted in April to pass on the news that peace preliminaries with France had been signed. Pelham's joy at seeing an end to the continental war was unreserved: 'I look upon it as almost a miraculous deliverance for this country and the Republic, considering the great and successful army of France, and the weak and unfortunate one of the Allies.' He gave Grenville a brief résumé of the terms and advised him to keep them secret: 'as they are not published, I think you had better not mention these particulars as having them from authority. We do not see the use in laying the preliminaries before Parliament.' The tone of this letter makes it clear that Grenville had quickly turned from rebel to trusted confidant after his move to the treasury. Nor is it difficult to see why. Pelham's efficient handling of administrative matters and constant search for economy in public expenditure mirrored Grenville's own approach to official business. In parliament Pelham shared a deep respect for the authority and tradition of the Commons. More important, they both also understood its constitutional forms and procedures with respect to the passage of legislation. Despite his illness, Grenville offered to return to London after the Easter recess: Pelham, however, assured him that his presence would not be necessary 'We have a Board complete at the Treasury', he wrote, and I look upon the sessions as over, our last Money Bill having passed the House of Commons last Friday.' Pelham's only worry for the summer was the king's imminent departure for Hanover. Yet it spoke for the ministry's strength that Pelham could comment: 'It is always best when he is amongst us; but as Peace is near at hand, I don't think his absence of equall ill consequences as it would have been, had not the preliminaries been signed.'[67]

To have received this clear acknowledgement of Pelham's approval must have pleased Grenville considerably after the months of trouble at the admiralty. The autumn brought further joy when Grenville succeeded in winning the hand of Elizabeth Wyndham, granddaughter of the Duke of Somerset. The marriage took place in May 1749 heralding

[67] The quotations are taken from Pelham's letter to Grenville on 30 April 1748, *Grenville Papers*, i. 74–5.

twenty years of happy family life until his wife's death in December 1769. The match also led Grenville into a political tie with his future wife's brothers. Charles, Earl of Egremont and Percy Wyndham O'Brien, later created Earl of Thomond in the Irish peerage. This connection proved extremely valuable in the 1760s when Grenville broke with Pitt and other members of his family to fill the highest offices of state. The only disappointment in preparations for the wedding came in December 1748 with the death of the Duke of Somerset. Contrary to expectation, his granddaughter received a small bequest of £100 per annum. This sum, as Walpole observed, 'is just such a legacy as you would give to a house keeper to prevent her going into service again'.[68] Nothing concealed the slight, and Somerset intended it as punishment for marrying into a family he held in contempt. It made no difference to the couple's happiness. Grenville secured a devoted wife and astute political companion: 'she was the first prize in the marriage lottery of our century', commented one of his friends in 1765.[69]

Grenville's absence from Westminster lasted until January 1749. Pelham's success in signing a peace with France brought its own problems in the House of Commons. Gaining approval for the preliminaries proved straightforward. Ministers presented the treaty to the Commons on 17 January and during the next two weeks easily defeated opposition motions respecting papers and correspondence used in framing the articles of peace. In ending the continental war, however, Pelham had left the government with the task of redeploying military and naval personnel, now no longer needed to keep the forces at wartime strength. To complicate matters further, ministers held different views on how this should be done, offering a perfect issue for the opposition to sink its teeth into. The army presented the initial problem. That a reduction was desirable found agreement in government circles. What caused dissension were the conditions under which the peacetime force would serve. The first sign of trouble came after the committal of the Mutiny Bill on 10 February. The committee appointed sat to thrash out clauses of the

[68] *Letters*, ii. 352.
[69] H.M.C. Lothian, p. 259.

bill for three and a half weeks, and it proved a rough ride. The first point at issue concerned an opposition move to exclude officers on half-pay from being subject to military law: 'It was feared that such a regulation would increase the power of the Crown, by causing those who were affected by it to use their influence in favour of ministers.'[70] Fortunately for the government legal witnesses testified that half-pay officers were already amenable to martial law and the amendment was defeated. As Walpole pointed out, however, 'it started up so formidable an Opposition as to divide 137 against 203'.[71] Open disagreements over other clauses in the bill were just as embarrassing for the ministry. In a bitter fight over a proviso to prevent the revision of a sentence by a court martial, after the party was once legally acquitted,[72] Grenville found himself not only at odds with Pelham and Pitt but also his brothers, James and Richard. The rancour caused by this incident is evident from Grenville's account of proceedings in the Commons on 7 March:

> When the clause was proposed in the Committee it was rejected. Mr. Pitt was absent from illness, and the question never having been agitated till the morning it was proposed, his opinion was not known. Lord Lyttelton, Mr. Campbell, and myself (then all three in the Treasury), voted against Mr. Pelham for the proviso, with many other persons in office, and as the rejection of it made a great deal of noise, the Court thought it necessary to bring in a clause themselves to restrain the revision to once only. When the report was made some time after, Mr. Pitt came down to the House and made a speech on purpose to declare his disapprobation of the proviso that had been offered, and to treat with slight the conscientious opinions of those who had voted for it? to which, however offensive it was, no reply was made, either by Lord Lyttelton or myself.

Grenville and the government were spared further blushes by the actions of Lord Egmont, leader of the opposition who 'walked out of the House on the Report, after an undignified display of temper, when Fox vigorously opposed an amendment which tended to weaken the oath of secrecy relating to the vote of individual members of courtsmartial'.[73]

[70] Ilchester, *Henry Fox, First Lord Holland*, i. 154.

[71] *Letters*, ii. 360.

[72] This quotation and the one that follows are taken from entries in Mrs Grenville's diary, *Grenville Papers*, i. 426–7.

[73] Ilchester, *Henry Fox, First Lord Holland* i. 155.

In the final division on the Mutiny Bill that day, the government's majority increased to fifty-eight votes, an improvement of twenty-two votes on the previous motion.[74]

Some months passed before ill feeling created by this incident subsided. In June 1749 George Dodington resigned his position as treasurer of the navy to join the Prince of Wales in opposition, and Grenville believed that he would be appointed in his place. The basis of this conviction lay in Pelham's promise three years earlier that Grenville's keenest rivals, Legge and Campbell, 'should not go out of the Treasury before me'.[75] It was not to be. Legge took Dodington's post and left Grenville bemoaning his lot. In view of his recent marriage this is understandable, for the job carried the handsome salary of £2,000 per annum. Yet what grieved Grenville most about the episode was the lack of support from his family and colleagues in his claim to promotion. 'I was sensible of this breach of promise', runs the note in Mrs Grenville's diary, 'but the situation Mr. Pitt was then in must have occasioned a rupture with him, and probably difficulties with my brothers (as none of them expressed any readiness to support my pretensions), had I asserted my claim.'[76] The disagreements over the Mutiny Bill, and Grenville's close relationship with Pelham had obviously cut deep into the family's solidarity. Grenville's pill of disappointment was sweetened some weeks later by a generous legacy from Lord Cobham who died on 13 September 1749. His death also signalled fresh honours for the family. Through Grenville's good offices with Pelham, his mother became Countess Temple with the remainder of the title passing to Richard on her death. This occurred three years later and with it the estates at Stowe and Wotton devolved on Richard, now Earl Temple, and George respectively. Grenville never owned Wotton, but leased the house from his brother for an annual rent in 1753 of £10. Grenville always relied on his brother for his return to parliament too, and this lack of personal security helps to explain much of George's obsession with gaining promotion and the higher salaries that went with it. More often than not, when these

[74] *Commons Journals*, xxv. 773–4.
[75] *Grenville Papers*, i. 427.
[76] Ibid.

ambitions were thwarted Grenville turned on his family and friends who did little in these years to repay his loyal service in their cause.

Grenville failed to attend the opening of the new session of parliament on 16 November 1749. His wife fell ill during the autumn and he spent the next nine months in Bath as she convalesced. Fortunately for the government his presence was not required. The Address passed with surprisingly little opposition. Egmont made 'a violent, and very injudicious speech' but the Address was voted without a division.[77] The government felt relieved at this turn of events because the opposition had been planning strategy for the coming session and Pelham was always wary of any threat to his position in the Commons.[78] His preparedness proved well founded after the Christmas recess when the opposition threw its forces into battling the annual vote of the Mutiny Bill. The targets were the same clauses, concerning revision of courts martial and oaths of secrecy, that troubled ministers the previous year. All went well for the ministry in the end because the opposition seemed unable to exploit the division of opinion within the government. Pitt sent Grenville two reports of debates on the Mutiny Bill in late January, and described opposition efforts to amend the legislation with contempt. If there were differing views within the government he did not consider Egmont and his followers capable of turning them to the opposition's advantage.[79] On 29 January the government defeated an adjournment motion on the bill by 161 to 89 votes.[80] On the Report Egmont acknowledged the opposition's impotence on this topic by leading his followers from the House before a vote was taken.[81] Pelham experienced equal success the following week when the opposition launched an attack on the government's failure to monitor French compliance with the peace treaty signed at Aix-la-Chapelle in 1748. The point at issue, George Dodington told members,

[77] *Dodington Diary*, p. 25.

[78] This can be seen from the collection of documents in *Leicester House Politics*, pp. 180–92.

[79] *Grenville Papers*, i. 93–5.

[80] *Commons Journals*, xxv. 968.

[81] Walpole, *Letters*, ii. 424.

was the question of fortifications at Dunkirk which the French agreed to dismantle but had since done nothing to fulfil this commitment. He wished for the papers on this clause to be laid before the House and a decision reached on what line should be taken with the French in the light of subsequent events in Europe. This moderate request might have succeeded but for Egmont's rash decision to force the issue. Laying the papers before the House was not good enough, in his opinion, the French should be made to comply with the treaty immediately and a vote was taken to execute this article of peace in the treaty. Ministers, led by Pitt, had no difficulty in defeating this motion as being too precipitate.[82] In a division that emphasized Pelham's superiority for the remainder of the session, the government crushed the opposition motion by 242 to 115 votes.[83]

The opposition clearly lacked the numbers and issues to make any impression on the government's majority. Yet, danger to Pelham did exist and, as ever, it came from within. It took the form of a growing antipathy between Newcastle, secretary for the northern department and Bedford, who since 1748, had been secretary for the southern department. By the spring of 1750 relations between the two had deteriorated to the point where Newcastle recommended to Pelham that Bedford be dismissed. Indeed, at the beginning of the year Walpole commented that 'the two Secretaries are on the brink of declaring war'.[84] Their dispute had many facets. Initially Bedford's protégé, Sandwich had upset Newcastle in 1748 when he was sent as special peace envoy to France. His inclination had been to report to Bedford rather than Newcastle, in defiance of Newcastle's wish to retain tight overall control of the negotiation. Newcastle was undoubtedly jealous, too, of Bedford's close relationship with the king and his favourite son, William, Duke of Cumberland. Of further annoyance to Newcastle in the years 1748–50 was Bedford's interference in some delicate negotiations with other European powers concerning the election of the King of the Romans. Newcastle intended that the Austrian nominee, Archduke Joseph should be elected

[82] *Dodington Diary*, pp. 45–6.
[83] *Commons Journals*, xxv. 977.
[84] *Letters*, ii. 424.

and directed his negotiations to outflanking the French before they could produce a candidate of their own. Throughout Bedford proved an unwilling partner in this plan. He did not believe Newcastle could dupe the French and opposed his plan on the grounds that it meant granting subsidies to European powers in a time of peace. To make matters worse, he began to formulate a policy for combating French aggression in North America without regard to Newcastle's delicate negotiations.[85] Members of parliament first became aware of their disagreements in January 1750, through two quite unrelated issues. The first concerned the election of Bedford's son-in-law, Lord Trentham, for the seat of Westminster. In late 1749, Trentham gained promotion to the admiralty board and was thus obliged to seek re-election to the House of Commons. The Westminster constituency was extremely difficult to control and the opposition sensed an opportunity to defeat the government candidate with their own nominee, Sir George Vandeput. At the poll they seemed to have failed, but such was the revulsion at Bedford's bullying tactics during the election the voters decided to petition the House, calling for a poll scrutiny. Much to Bedford's chagrin, Pelham allowed the petition and the scrutiny to take place, and refused to use the government's majority in the Commons to end discussion of the affair. Debates on the election dragged on not only during this session but also the following year as well, and Bedford soon saw Newcastle's hand behind an obvious attempt to discredit him.[86] The second issue concerned a turnpike bill sponsored by Bedford and presented to the Commons on 29 January. The bill proposed the construction of a road from Westgate in Bedfordshire to Market Harborough in Northamptonshire, and, from its inception, had aroused considerable opposition from the towns and villages lying on its route. Bedford hoped this problem could be overcome by Pelham flexing his muscles in the Commons and at first this seemed to be the plan. In a very full House on 27 January the bill was committed by 197 to 186 votes.[87] Between this

[85] Newcastle's foreign policies during this period are examined in depth by Reed Browning, 'The Duke of Newcastle and the Imperial Election Plan, 1749–1754', *J.B.S.* vol. II. 1 (1967), pp. 28–47.

[86] The election is explained in detail in Sedgwick, *Commons*, i. 286–7.

[87] *Commons Journals*, xxv. 968.

narrow squeak and the report on 13 February, however, the Pelhams had second thoughts. Faced with a number of petitions protesting at the road and the likelihood of the opposition taking up the issue, they decided to concede the point. In a predetermined manoeuvre they allowed the bill to be defeated by 208 to 154 votes.[88] 'The Pelhams, who lent their own persons to him, had set up the Duke of Grafton, to list their own dependents under against their rival.'[89]

This dispute entered its final phase during the remaining months of the year. Newcastle spent the summer of 1750 in Hanover with the king without diminishing his hostility to Bedford. Pelham shared some of his brother's distrust of the duke but stopped short of advocating his dismissal for fear of offending the king. In the summer he attempted to solve the problem by offering Bedford the prestigious post of master of the horse, which became vacant on the death of the Duke of Richmond. Bedford saw through this ruse to remove him from the cabinet, and only agreed on the condition that he could name Sandwich as his successor. This was anathema to the Pelhams who both detested Sandwich, and at this point peacemaking came to an end. Bedford's obduracy convinced Pelham that he must be removed, and in November he found the excuse to pursue this goal. For over a year the cabinet had been discussing the question of the naval establishment in peacetime. The Pelhams felt that a reduction in the number of seamen, similar to that achieved for the army, was necessary to reduce public expenditure and avoid a mushrooming navy debt. This proposal met stiff opposition from Sandwich at the admiralty who wished to keep the navy at near war-time strength to combat the threat of French aggression in North America. There seemed little doubt that Sandwich would lose the argument. In September Pelham told the Earl of Marchmont that: 'Quiet is what we want; economy is necessary; but the one cannot be had without the other.'[90] In November Pelham did carry his case in cabinet but only after a good degree of bloodletting.

[88] *Commons Journals*, xxv. 993.

[89] Walpole *Letters*, ii. 427. In this letter Walpole gives the minority as 52 instead of 54.

[90] *Marchmont Papers*, ii. 387.

After one particularly stormy meeting that included discussion of the navy estimates as well as vacant places in the administration, the cabinet broke up in uproar. Lord Egmont recorded the scene in his diary on 24 November:

Rumours of great differences in the Cabinet. The Duke of Cumberland and Bedford, Fox and Legge against Pelham, Duke of Newcastle, Pitt, Lyttelton, Murray etc. . . . that the Duke of Newcastle had sounded who would stand by him – the Chancellor wanting Anson for the first Commissioner of the Admiralty playing a double part – Pelham uncertain, puzzled and double – but Pitt, Lyttelton, Murray etc. standing firm.[91]

Pitt and Lyttelton did well to stand behind Newcastle. He and Pelham were in no danger of losing the battle over a reduction in the number of seamen with either Bedford or Sandwich. In a meeting of the Privy Council on 21 November, Pelham set the limit of seamen at 8,000 and preparations for presenting the measure before the Commons began.[92] It appeared on 22 January 1751 in the Committee of Supply, when Lord Barrington moved that the number of seamen for the coming year should not exceed the 8,000 limit agreed in November.[93] No one spoke up for keeping the navy at wartime strength but Egmont, Nugent, and Potter declared in favour of the normal peacetime establishment of 10,000 seamen. A short debate then followed in which Pitt observed without prior warning to his colleagues 'that if the motion had been made for ten thousand, he should have preferred the greater number'. Pelham tried to gloss over this difference of opinion but to no avail. The opposition saw an easy opportunity to embarrass the government and forced a motion for the higher number of seamen mentioned by Pitt. With reluctance Pelham and Pitt were obliged to stand by their statements and vote on opposite sides: 'the eight thousand were voted 167 to 107; only Pitt, Lyttelton, the three Grenvilles, Colonel Conway, and eight more, going over to the minority.'

Walpole made a great deal of this incident believing it to be the precursor to a campaign by the Cobhams to elevate

[91] *Leicester House Politics*, p. 193.
[92] *Dodington Diary*, p. 91.
[93] Walpole, *Memoirs of the Reign of George II*, i. 12. The account and division that follow are taken from this source too.

Pitt to the post of secretary of state, if and when Bedford departed the scene.[94] He supported this accusation with a further charge that Pitt and the Grenvilles had sought a reconciliation with the Prince of Wales in the autumn to put pressure on Pelham for action over Bedford. Two factors weaken this argument. First, it is clear from Egmont's papers that the overtures to the Prince were a joint venture between Pitt and Newcastle.[95] If this is the case, it seems more likely that isolating Bedford rather than promotion was the main motive behind Pitt's action. Helping the Pelhams was always in Pitt's best interest, and Newcastle's plan to remove Bedford depended on cutting off the supply of potential allies to the duke, especially those like the prince who had a voice at Westminster. Second, in the debate on the Address on 17 January, Pitt reiterated the commitment to Pelham's administration given in the cabinet during November.[96] The disagreement in the House on 22 January stemmed from a piece of ignorant behaviour by Pitt which he immediately regretted. When the report on the reduction to 8,000 seamen came before the House seven days later, Pitt, along with Grenville and Lyttelton, apologized to Pelham for contradicting his wishes.[97] This did little for their respect and popularity with the king or members of parliament, but consolation lay in the fact that on this occasion they entered the government lobby, securing a majority of 189 to 106 votes.[98]

Grenville's first involvement in a parliamentary debate for over twelve months thus passed inconspicuously. It is hardly surprising therefore that he and Pitt lay low for the remaining weeks of the session. The opposition tried to raise campaigns on several issues during February and March but made little headway. In proposing a land tax of 3 shillings on 13 February, for example, Pelham emphasized that no matter what differences he had with Pitt and Bedford, the government's position remained secure. On the Report on the land tax five days later the opposition suffered a crushing defeat by 229 to 28 votes,[99] prompting Walpole to comment

[94] *Letters*, iii. 32–3.
[95] *Leicester House Politics*, p. 193.
[96] Walpole, *Memoirs of the Reign of George II*, i. 8–9.
[97] Walpole, *Letters*, iii. 32.
[98] *Commons Journals*, xxvi. 22.
[99] Ibid., p. 49.

that the minority 'could not conjure up a spirited division now on the most popular points if they were not new, they would scarce furnish a debate'.[100] In debates on the Mutiny Bill the opposition experienced better success; though in truth this happened more by accident than design. In the Committee of the Whole House on the Mutiny Bill of 20 February, Sir Henry Erskine, the opposition MP for Ayr Boroughs, launched an attack on the well-worn issue of revising the decisions of courts martial.[101] During his speech he complained of the wide-sweeping powers vested in general officers on courts martial, citing in support his own trial ten years earlier in Minorca under General Anstruther, now MP for Anstruther Easter Boroughs. This veiled accusation led to a protracted dispute in which the opposition tried to build on Erskine's statement and what appeared as a real cause for redress. It took the form of a motion by George Townsend on 4 March requesting all the papers relevant to the trial in Minorca of ten years previous to be laid before the House. After a short debate the ministry granted Townsend's wish, only to be confronted the following day with a petition from a Minorchese, Don Juan Compagni, requesting compensation for legal expenses accrued during Anstruther's term as governor of Minorca from 1733 to 1747. It appeared that Anstruther had condemned Compagni for crimes he did not commit, and the Minorchese received no compensation after clearing his name in England. This became the cue for taking sides, with opposition supporting Erskine and the government defending Anstruther. Pitt and Grenville, however, found it impossible to justify Anstruther's known brutality in Minorca and supported the petition, 'on the fitness of granting two or three thousand pounds to a poor man oppressed by military law'.[102] Pitt then went a step further by asking for a parliamentary inquiry to settle the matter once and for all. Pelham did not go along with this line of argument. He told the House that an attempt would be made to have Anstruther tried by court martial, but he could not support a move to have the treasury paying legal expenses.

[100] *Memoirs of the Reign of George II*, i. 32–3.
[101] The description that follows is taken from Sedgwick, *Commons*, i. 417–18.
[102] Walpole, *Memoirs of the Reign of George II*, i. 58.

Finally on 5 March, Pelham ended all talk of the petition by moving for the orders of the day. The opposition sought a division and went down by 97 to 58 votes.[103] In the minority were Pitt, Conway, and the three Grenvilles. It is unlikely that Pelham felt rancour at this exchange. The motives for supporting the petition were justified, and there could have been little danger to the government as the case appeared so far removed from discussion of the Mutiny Bill. In fact Walpole records that Pitt tried in vain to prevent a division when he saw that Pelham would be on the opposite side.[104] Both Pitt and Grenville came out of the episode with some merit, when it became clear that a parliamentary inquiry could be the only means of settling the matter. As Pelham told the House on 24 April, the Act of Grace passed in 1747 forbade Anstruther's case being retried before a court martial.

This brief surge of activity proved to be the opposition's heyday. On 20 March 1751 Frederick, Prince of Wales, died, leaving Egmont in charge of a leaderless and aimless flock. The heir to the throne, Prince George, was then only thirteen years old. The Pelhams, strengthened by the confusion that followed the Prince's death, carried through the long-desired removal of Bedford. The excuse for acting came from another disagreement over policy in North America. Since the new year, Halifax, president of the board of trade, had been recommending that a naval force be sent to Nova Scotia to counteract French aggression in the region. Sandwich at the admiralty and Bedford in the southern department disagreed with this view and resisted complying with his request. The issue then went back and forth for over two months, until, in June, Pelham gained the king's permission to dismiss the unpopular Sandwich and appoint Anson in his place. It was a clever ploy, for the Pelhams knew that Bedford would feel duty bound to resign in a gesture of loyalty to his protégé. This he did on 18 June, and with Bedford gone, Cumberland's influence over the king diminished accordingly. In the changes of personnel that followed the Pelhams finally established a united ministry within to match their power at Westminster. Pitt and

[103] *Commons Journals*, xxvi. 91.
[104] *Memoirs of the Reign of George II*, i. 60.

Grenville remained in place, and it has been said by the family's biographer that Newcastle erred in not promoting them at this juncture.[105] Events over the next three years to Pelham's death do not bear this out. Pitt and Grenville never entered the minority lobby again or openly disagreed with their leader. The reasons for this are not difficult to find. First, they needed Pelham more than he needed the remaining Cobhams. The dominant position of the ministry in parliament acted as an instrument of discipline on all those thinking of misconduct in the House. Second, Bedford's departure gave Pitt and the other Cobhams a much greater voice in government decision-making. The early proof of this came after the opening of the new session on 14 November 1751. The week after the Address Lord Barrington moved to increase the number of seamen to 10,000 for the ensuing year, making no secret of the fact that the change had been prompted by pressure from certain quarters in the government.[106] The motion passed without dissent. Thus, if the Pelhams reacted to all their requests like this, what point was there in Pitt and Grenville causing trouble?

The most perceptive judgement on these changes came from Walpole who merely mourned the passing of opposition at Westminster: 'What I believe was never yet told of an English Parliament, that it is so unanimous, that we are not likely to have one division this session – nay, I think not a debate. On the Address, Sir John Cotton alone said a few words against a few words of it.'[107] His prophecy almost came to fulfilment. Only a short bout of opposition to a subsidy treaty with Saxony in January 1752 disturbed Pelham's calm that session. Excitement arose over the issue because the Duke of Bedford appeared to be moving towards an opposition alliance with Egmont. Hopes of achieving anything in this area of foreign policy perished, however, as quickly as they were raised. Pelham brought up the Saxon subsidy in the Committee of Supply on 22 January, and the proposal met immediate opposition from Horatio Walpole, Sir Walter Blackett, William Beckford, Lord Strange, and

[105] Wiggin, *Faction of Cousins*, p. 143.
[106] Walpole, *Memoirs of the Reign of George II*, i. 211.
[107] *Letters*, iii. 75.

Grenville's elder brother, Richard, now Lord Cobham.[108] These speeches made little impression on the House despite the fact that there was general resentment at such expenditure in peacetime. In part this was due to Walpole's speech in which he recited several criticisms of the treaty only to conclude by saying he would vote with the government. But, more important, Pitt and Grenville refused to join Cobham's opposition to the treaty. The eldest Grenville now immensely wealthy and holding no official post could lose nothing by defying the ministry. Throughout the remaining years of his life he gave frequent demonstration of opposing the government for its own sake. Yet unlike his uncle, he held no political sway over the family and Pitt and Grenville declined to support his erratic behaviour on this and many other subsequent occasions. Even Cobham himself saw the futility of opposition on the Saxon issue by walking out before the division: the House approved the treaty by an overwhelming majority of 236 to 54 votes. On the Report the following day no division took place at all. On 28 January Bedford made a token effort to expose the treaty in the Lords but it too was doomed to failure. A note in Dodington's diary sums up the debate: '28. Tuesday. Duke of Bedford attacks the subsidiary Treaty with Saxony ALONE.'[109] A third attempt to salvage something from this issue was made by the opposition in the Commons on 29 January. A group of Tories moved an Address to the king not to enter into subsidy treaties in times of peace. It fared no better than its predecessors and went down to defeat by 182 to 52 votes.[110] Thus, to all intents and purposes, ended the parliamentary session of 1752.

Grenville took no part in any of the debates that occurred in the remaining weeks of that short session. He spent most of his time that year alternating between official duties in London and directing improvements to the Wotton estate

[108] The account of this debate and the division are taken from Walpole, *Memoirs of the Reign of George II*, i. 242–3.

[109] *Dodington Diary*, p. 143. The editor's footnote on this page is inaccurate: from the division figures it seems clear that the House was not 'singularly thin' during the debate.

[110] *Commons Journals*, xxvi. 411.

in Buckinghamshire. The next session of parliament opened on 11 January 1753 and from the beginning it appeared that Pelham's Indian summer would continue. The Address passed without serious opposition, and despite a renewed effort on Egmont's part to censure the army estimates on 26 January, these too passed with a division.[111] Bedford's flirtation with opposition had done little to repair the damage of Frederick's death: as Professor Foord pointed out in his survey of the period, from Bedford's resignation in 1751 to Pelham's death in 1754 the minority vote never even approached 100.[112] It is a tribute to Pelham's achievement as first minister that the most contentious issue before parliament that session concerned a clandestine marriages bill, introduced by Lord Bath in the Lords during May. It rose to prominence simply because Fox, the secretary of war, decided to oppose its passage and became involved in a bitter dispute with the lord chancellor, Hardwicke. There was never any danger for Pelham in the lobbies, where every vote went the government's way.[113] Grenville's sole recorded contribution to parliamentary business in the 1753 session also came in May when he supported a bill, on the third reading, for 'taking account of the total number of people, and of the total number of marriages, births and deaths; and also of the total number of poor receiving alms . . .'.[114] In a very thin House the opposition did not challenge the bill outright, but, led by Matthew Ridley and Thomas Pitt, demanded amendments aimed at preventing gross inaccuracies and unforeseen expense. Grenville's speech acknowledged the defects of the legislation, while maintaining that they did not detract from the value of such a register in planning future domestic legislation. This sensible view seemed to reflect the mood of the House, for the bill was carried by 57 to 17 votes.[115] The Lords took a different line, however, throwing the bill out on its second reading in the Upper House. Grenville had chosen to endorse one of the few policies where Pelham failed to have his own way.

[111] Ilchester, *Henry Fox, First Lord Holland*, i. 183–4.
[112] *His Majesty's Opposition*, p. 280.
[113] Walpole, *Memoirs of the Reign of George II*, i. 336–49.
[114] *Parliamentary History*, xiv. 1330–65, gives a brief description of the debate.
[115] *Commons Journals*, xxvi. 810.

Grenville's summer holiday lasted until 15 November when the final session of the 1747 parliament opened. The House was well attended in expectation of an opposition campaign concerning a piece of legislation relating to the naturalization of British Jews passed in May of that year. Its progress through the House had aroused a few murmurs of discontent but the division of 95 to 16 votes for the committal on 8 May bears witness to a relatively trouble-free career for the Act.[116] During the summer recess, however, a marriage of interest between the Duke of Bedford's money and an opportunist newspaper editor, James Ralph, spelt trouble for it. Together they produced a popular news-sheet, *The Protestor*, and agitation against the Jewish Naturalization Act became its principal function. In part this reflected Bedford's desire to exploit any issue that would embarrass the government and give the opposition a lead at Westminster. But just as important were the other groups within society who found fault with the legistlation. These ranged from the merchants in the City of London, who had originally petitioned against the bill, to the country gentry and their spiritual ally the High Church of England. Some of their attacks on the legislation were blatantly anti-semitic, but their basic motive for opposition lay in self-interest: 'the Jew Bill ultimately rested upon basic principles of religious and economic "liberalism" which were incompatible both with the nationalist and High Church opinions of the Tory remnant and with the restrictionist economic views of the City merchants.'[117] Pelham took this opposition seriously because of its threat to the government's fortunes in the coming general election. By July 1753 the ministry had already taken stock of the popular wave of protest and decided on outright repeal of the Act, if the tide of opposition did not recede. In the event no decline took place and Newcastle presented the repeal to the House of Lords on 15 November, with Pelham introducing the measure to the Commons eight days later. In its case for repeal the government made no secret of the fact that principle would be sacrificed for political gain. In the Commons the spectacle

[116] *Commons Journals*, xxvi. 809.

[117] Perry, *Public Opinions, Propaganda, and Politics in Eighteenth-Century England*, p. 70.

of MPs supporting the repeal who obviously felt strongly that the Act should remain on the statute-book represented a sorry sight.[118] Their only consolation, as Sir George Lyttelton pointed out, was the fact that no Jews would have taken advantage of the Act in the present climate of opinion so its loss would not be felt. Grenville supported repeal but refrained from speaking in any of the debates. The bitterness he felt at the government's action and the shameful aspects of the protest were summed up by Pitt in his speech to the Commons on 27 November during a Committee of the Whole House on the repeal bill: 'I must still think that the law passed last session in favour of the Jews was in itself right, and I shall now agree to the repeal of it, merely out of complaisance to that enthusiastic spirit that has taken hold of the people; but then I am for letting them know why I do so.' In his defence Pelham could claim that the government did stop short of complete surrender on the issue. On 4 December the Tory, Lord Harley, moved that a similar Act affecting the colonies passed in 1740 be repealed, but in this instance, the government refused even to consider the matter. 'Repeal', Pelham said, 'would be to tell the people, we will repeal this law, not because it has, but because it ought to have made you uneasy.'[119] Doubtless on this occasion Grenville felt no pangs of conscience when he became part of the huge majority that rejected the motion.[120] Nevertheless, of the old Cobham group, only Richard Grenville, recently raised in the peerage from Viscount Cobham to Earl Temple, came out of these debates well – attacking, throughout, the government's hypocritical actions in the face of an indefensible campaign.

Grenville's last leading part in parliamentary business under Pelham took place on 12 December 1753, when he acted as treasury spokesman in a debate on the importation of French cloth into Britain. Sir John Barnard, the wealthy financier, supported by William Beckford, moved that restrictions affecting the 'weaving and importation of

[118] This account of the debate and the quotation that follows are taken from *Parliamentary History*, xv. 119–55.

[119] Walpole, *Memoirs of the Reign of George II*, i. 366.

[120] *Commons Journals*, xxvi. 861. The figures were 208 to 88 votes.

Cambricks and French lawns' be repealed.[121] Their arguments concentrated on the futility of such wartime restrictions being enforced during a period of peaceful trading, when a market for such goods existed in Britain. It seemed a fair point but the government refused to give an inch. In a firm reply, Grenville said the prohibitions would remain, the government wished to protect British manufacturers and had no intention of encouraging those in France. It is likely that knowledge of French aggression in North America and the East Indies prompted this strong stand, and there is no doubt that it found favour with those MPs present. Despite Barnard's strong case for repeal, the motion fell to defeat at the first hurdle without a division. After attending to this official business Grenville retired to Wotton for the Christmas holidays, but he did not return in January to take part in the debate at the start of the session. The reasons for his absence remain a mystery. Only a note in Dodington's diary on 24 January 1754 gives some clue, where it is recorded that complaints were made at court of 'G. Grenville's insolence in refusing to come to town'.[122] If Grenville was sulking or unhappy about ministerial policy, as this implies, it is difficult to see which issue upset him. Two topics that session caused debate in the House: a move to repeal the bribery oath in elections, and a bill for subjecting to military law the troops going to the East Indies, and Grenville had no quarrel with Pelham on either. Indeed, it is probable that Pelham knew of or even gave permission for Grenville's absence in the full knowledge that his presence would not be needed. The last division of note before Pelham's untimely death on 6 March went the government's way by the emphatic margin of 245 to 50 votes.[123]

Pelham's demise brought a sudden end to a settled and rewarding period in Grenville's political life. From a bad start at the admiralty he achieved a good working relationship with Pelham after his promotion to the senior board. Had he survived, Grenville's rise up the ministerial ladder would have been a predictable and uncomplicated progression.

[121] The description that follows is taken from *Parliamentary History*, xv. 163–91.

[122] *Dodington Diary*, p. 247.

[123] *Commons Journals*, xxvi. 961.

His future now looked less promising. The king put the picture in perspective when he commented woefully on hearing of Pelham's death 'I shall now have no more peace.'[124] The era of political stability, last enjoyed in the era of Walpole, perished with Pelham, and in the struggle for supremacy that followed Grenville and his powerful allies could not guarantee that they would come out on top.

[124] Walpole, *Memoirs of the Reign of George II*, i. 378.

Chapter III

Coming of Age, 1754–1760

The death of Henry Pelham represented a serious setback to Grenville's political career. His promotion from the treasury board had seemed assured with the passage of time, for in Pelham he could not have found a more powerful or respected ally on whom to pin his hopes of future preferment. Pelham's sudden demise now threw such promise into doubt, and in the months that followed Grenville felt the loss keenly. After March 1754, only the alliance with Pitt and the other Cobhamites offered a real chance of promotion, and by the end of the year Grenville was professing little faith in that. This disillusionment began with the ministerial changes that took place after Pelham's death. In the power struggle that ensued, the Duke of Newcastle, supported by the king and the Earl of Hardwicke quickly exerted control over the remnant of Pelham's administration, reducing Grenville, Pitt and the rest of the Cobhamite group to the role of passive bystanders. The ease by which Newcastle executed this manoeuvre was due in part to Pitt's confinement at Bath with gout, but it also owed something to a clear breach of trust in the Cobhamite ranks. In a skilful piece of negotiation by the court, Sir George Lyttelton was tempted to push himself forward as the unofficial spokesman of the group in Pitt's absence. Dazzled by the offer of the post of cofferer and a vague promise of a peerage, Lyttelton failed to consult the tender sensibilities of his colleagues, and during the negotiations following Pelham's death committed the whole group into accepting whatever arrangements Newcastle saw fit to make. In the event they turned out to be unsatisfactory: Pitt remained paymaster and Grenville accepted the post of treasurer of the navy with a salary of £2,000 per annum, and they reacted with characteristic bluntness to Lyttelton's presumptuous behaviour. It was spelt out in no uncertain terms that in future they 'desired to speak for ourselves whenever the occasion should require it'.[1]

[1] *Grenville Papers*, i. 430.

This was not the end of the matter for Grenville. During the next six months it became clear that his personal disenchantment with appointments in the administration did not stop with Lyttelton's misguided actions. After Newcastle's arrangements were finalized, Grenville began to question Pitt's behaviour as leader of the group during March. In particular, Grenville doubted whether his own claims for high office, with cabinet rank if possible, had been canvassed in any serious way with Newcastle and Hardwicke. In adopting this view Grenville could be criticized for harbouring delusions of grandeur, for his stature in the Commons or at Westminster in general hardly warranted such bold aspirations. Be that as it may, Grenville had certainly been led to believe that an important post would come his way. In fact on 7 March Pitt told his brother that 'George Grenville may be offered Secretary at War'.[2] Moreover, in promoting the modest talents of Legge and Robinson to the posts of chancellor of the exchequer and southern secretary respectively, the Court severely undermined its case against Grenville's claim for advancement. His abilities as an administrator and debater were at the very least equal to both these men. Was Pitt to blame, therefore, for failing to secure the desired promotion? In this case it would seem not. Pitt pressed Hardwicke as hard as he could but found that Newcastle had outwitted him on every front, including that of patronage. Hardwicke told Pitt on 2 April that treasurer of the navy was the best that could be done in the circumstances. 'I agree that this falls short of the mark', he wrote, 'but it gives encouragement.'[3] Pitt accepted this realistic explanation of the situation, and tried to be kind to Grenville when passing on the news by assuring him that the office would give 'weight in Parliament' to his claims for promotion in the next round of ministerial appointments.[4] This held no water with Grenville. His resentment at what he believed was an act of treachery burned all summer long and burst forth in his account of the wedding between his sister Hester and Pitt during the

[2] *Grenville Papers*, i. 111.

[3] *Lyttelton Memoirs*, ii. 459. For further information on the events of March to September 1754 see Clark, *The Dynamics of Change*, pp. 55–86.

[4] *Grenville Papers*, i. 119 (6 Apr. 1754).

autumn of that year. 'I did not imagine', he noted bitterly, 'that my behaviour [introducing Hester to Pitt] upon this occasion was to bring an enemy instead of a friend into our family.'[5] It is ironic that of all the occasions when Grenville had grounds to accuse Pitt of not protecting his interest, he should raise a storm over an incident in which his brother-in-law appears to have behaved impeccably.

In view of what had passed between them before 1754 it is not surprising that Pitt should take the full blast of Grenville's frustration at Pelham's death and the chance of ministerial promotion that perished with him. Grenville now had to decide how far his personal feelings should be allowed to affect his political behaviour. In the short term the answer was, hardly at all. Not for the first time Grenville found himself trapped between the devil, Pitt, and the deep blue sea, in this case Newcastle, whose attitude to Grenville was barely civil. Grenville therefore decided to leave matters as they were and make the best of his new office. He did not interfere with Pitt's plans, begun on the news of Pelham's death, for a strategy that would govern the conduct of the Cobham group over the coming months. It proved a wise step, for the strategy, as explained to Temple on 11 March, amounted to little in the way of positive action. They were all to remain in office, Pitt advised, and wait for the right moment to seek their revenge.[6] The only part of the plan that appealed to Grenville was a suggestion that the group throw its weight behind an alliance with the Princess of Wales. Lines of communication had been opened before Pelham's death, and it seemed sensible to Grenville on two counts to strengthen the bond. First, from the practical view, that in a short time Prince George would attain his majority, and when his elderly and weary grandfather died they would be in a favoured position at the start of a new reign. Second, such an alliance would be of immediate value in undermining the influence at Court of Cumberland, Bedford, and even Fox, whose hold over Leicester House had been so powerful during Frederick's lifetime. The princess became a willing ally in this realignment because she hated

[5] *Grenville Papers*, i. 431–2.
[6] Ibid., p. 112.

Cumberland, and had no intention of allowing her favourite son to fall under his tutelage. Grenville himself found further reward in cultivating this connection in that it eventually led him into a personal alliance with the Earl of Bute, Prince George's tutor and confidant. The emergence of Bute in Leicester House politics gave Grenville the hope of deliverance from Pitt that he had lost on Pelham's death.[7]

Nothing occurred in the remaining months of 1754 to lessen Grenville's discontent with his position in the ministry. On 21 June Grenville was nominated a privy councillor but in the most unsatisfying way. As Dodington wrote in his diary 'the Treasurer of the Navy, and the Cofferer were sworn of the Council: it has hung these three months since they had their offices, before the king could be brought to admit them.'[8] It is little wonder that earlier in the month Grenville told his colleague the Earl of Hillsborough, that he had expected 'very different treatment, consideration, and communication, from his rank in the House of Commons'.[9] He soldiered on through the autumn bemoaning his lot, and, as the parliamentary session approached, resigned himself to supporting Pitt's policy of acquiescence in Newcastle's desires. This philosophical attitude on Grenville's part certainly made sense in the light of the fact that the ministry was likely to have a majority of over 200 MPs: a clear tribute to Newcastle's management of the general election in the spring of 1754. Indeed, Grenville's own attempt to secure the return of a cousin, Temple West, against the government interest at Bishop's Castle failed miserably.[10] This experience, coupled with Newcastle's strength in both Houses, led Grenville to believe that making trouble at Westminster to signify Cobhamite disapproval of events of the spring would not be productive. He passed these doubts on to Pitt but his brother-in-law would hear nothing of caution and plotted a course in the opening days of the session that would lead him into conflict with his superiors.

[7] This alliance was sealed by the spring of 1755, see Ilchester, *Henry Fox, First Lord Holland*, i. 254.

[8] *Dodington Diary*, p. 283.

[9] Ibid., p. 279.

[10] Wiggin, *Faction of Cousins*, pp. 154–8.

Pitt intended, with the help of a disgruntled Fox, to undermine Newcastle's confidence by a personal attack on the new secretary of state Sir Thomas Robinson. Grenville believed that there was something cowardly in this plan. Robinson held high office on the basis of a long and distinguished diplomatic career, spent mainly in Germany. In no manner at all could he match Pitt or Fox in the Commons. Grenville therefore decided, as he put it, to dissuade Pitt 'from pushing things to extremities . . . as there was then no measure on foot to oppose'.[11] The plea to concentrate on measures rather than men fell on deaf ears, for Pitt wished to strike at Newcastle as soon as possible. Parliament opened on 14 November and in the debate on the Address it soon became clear that, even without a seasoned leader, Newcastle held an unassailable position in the Commons. Pitt sensed the mood of the House was with the ministry and for over a week remained silent. On 25 November, however, Pitt's patience ran out, and he and Fox delivered an impromptu character assassination of Robinson during a sitting of the Committee of Elections. As expected, they dealt mercilessly with Robinson, exposing his lack of experience in debate. But, as Grenville predicted, the tactic rebounded on them. Walpole sums up the day's events well: 'It was plain that Pitt and Fox were impatient of any superior; and as plain by the complexion and murmurs of the House in support of Sir Thomas Robinson, that the inclinations of the members favoured neither of them.'[12]

Grenville says of this incident that he 'joined with them, and gave them all the assistance I was able in the House of Commons'.[13] If this is true, it must have been very limited assistance. His name does not appear in any accounts of the debates in the session before the Christmas recess, and, in view of the fact that Grenville opposed this personalized attack, his support probably went no further than attendance in the House.

Grenville could not be faulted in his caution. This unpopular attempt to weaken Newcastle's position from within failed miserably, and was knocked firmly on the head in

[11] *Grenville Papers*, i. 431.
[12] *Memoirs of the Reign of George II*, i. 410.
[13] *Grenville Papers*, i. 431.

February 1755 when Newcastle ended Fox's flirtation with Pitt by offering him the place in the cabinet that he craved so dearly. Fox remained secretary at war but would now have a real say in policy, and places would be found for several of his friends. Grenville approved of this whole proceeding, for he had had little faith in an alliance with Fox. Pitt soon came to share this view, telling Fox on hearing the news of his promotion that it 'put an end to the intercourse that had subsisted between them'.[14] Grenville and Pitt were now out in the cold, 'at variance with the Administration, though still in office'. They required an immediate boost to their position and morale at Westminster, and it came in the shape of a formal alliance with the Princess of Wales at Leicester House. An informal agreement established the previous spring was now made a rallying cry for malcontents within the ministry. In his account of this public agreement Grenville makes no effort to obscure its importance to their political survival. He does point out that the princess endured the same harsh and unkind treatment at the hands of the king and his favourite son, Cumberland, suffered by the Cobhams, but this is little more than window dressing. The bargain was struck with Leicester House in a moment of desperate need, after Newcastle had again outwitted his internal foes. The announcement of their ties to the princess and the future king 'were given in the most public manner' and soon after the desired end was achieved when 'A party began to form itself under that standard'. From this point Grenville and Pitt survived in the ministry on borrowed time. The Earl of Waldegrave summarized their untenable position over the coming months with this wry observation:

> The substance of the treaty was, that Pitt and his friends should do their utmost to support the princess and her son; that they should oppose the duke, and raise a clamour against him, and as to the king, they were to submit to his government, provided he would govern as they directed him.[15]

This uneasy situation lasted some eight months. Newcastle spent the summer and autumn months attempting to bring

[14] Ibid., pp. 432–3. The two quotations that follow concerning these events are also taken from here.

[15] *Waldegrave Memoirs*, p. 39. See also Mckelvey, *George III and Lord Bute*, chap. 2.

the Cobhams into line. It proved a thankless task. Pitt continued during negotiations with Newcastle and the lord chancellor, Hardwicke, to make himself disagreeable in every way concerning policy and patronage. To judge from two lengthy accounts sent to Grenville and his brother Temple, it soon became clear that Pitt would only cease his opposition from within if Newcastle stepped down.[16] In view of Newcastle's strong position, Pitt's attitude appeared ill-judged, but two factors seem to have prompted this intransigence. First, Pitt obviously felt that the overblown alliance with the Princess had bolstered his position within the ministry and at Westminster in general; second, he wished to retain his independence from Newcastle in the hope of exploiting the government's conduct of foreign policy now the threat of war with France in Europe and North America was looming on the horizon. Since the autumn of 1754 a growing number of incidents between British settlers and French forces in North America, especially down the Ohio valley, had led to a general expectancy that conflict would ensue. More important, it would be a conflict that would affect British interests in Europe and the East as well. Pitt could find little fault with the ministry's initial response to the defence of its overseas possessions by deployment of naval forces. On the question of protecting Hanover, however, he did hope to raise a storm against Newcastle's subsidy treaties with friendly European powers. The basis of a parliamentary attack would be the well-trodden ground of constructing expensive alliances, in this case with Hesse-Cassel, to the benefit of Hanover and detriment of Britain. By the opening of parliament in November, Pitt would also be able to add weight to his cause by citing the disastrous efforts made by British naval forces under Boscawen to stop French reinforcements reaching America during the summer. Fortunately for Newcastle no one outside the Cobham group believed that the situation in September 1755 had yet developed into a serious threat to British possessions and interests overseas. In his concluding discussions with Pitt before the parliamentary session began, Newcastle could afford to be dismissive: 'it was said that the defection of

[16] *Grenville Papers*, i. 137-48. The quotation that follows is taken from here.

six or seven men could not create any disturbance to a measure [the Hesse-Cassel treaty] so popular.' The ultimatum had been delivered.

Grenville viewed these deliberations from the safety of Wotton, deliberately setting himself apart from Pitt's actions at this time. This coolness had nothing to do with policy, but concerned a personal grievance felt by Grenville at Pitt's decision in the summer to introduce Henry Bilson Legge, the chancellor of the exchequer, to Leicester House circles. Legge had risen to prominence in July 1755 for refusing to sign the treasury warrants allowing payment of the subsidies to Hesse-Cassel and other European allies without first laying the treaties before parliament. Newcastle, who had accompanied the king to Hanover for the summer, was forced to return early and deal with the situation. The subsidies were paid but Newcastle could not prevent Legge joining the Leicester House circle nor did he make any headway with Pitt in the negotiations of the autumn.[17] Grenville cared little for either Newcastle's dilemma or Pitt's line in the negotiations; his attention lay riveted to the fact that Pitt introduced Legge to Leicester House 'as the person the fittest to put at the head of the revenue, as Chancellor of the Exchequer, in the future reign'.[18] This struck right at the heart of assurances and promises given to Grenville by Pitt himself over the preceding twelve months and caused considerable antipathy to develop between the two. Indeed, it is worth quoting Grenville's account of Pitt's dealings with Legge to see how the relationship began foundering on a growing disillusion and distrust. Of the show of favour to Legge, he observed

> this destination made by Mr. Pitt in favour of a new acquaintance, without my knowledge or participation, though he could not forget what had passed not long before, upon the promotion of that gentleman over my head, was so contrary to those repeated professions of his wishes to see me at the head of the House of Commons, that it gave me a proof how little reason I had to depend upon them.

Grenville responded to this slight by taking a more independent line at Westminster when the new session opened

[17] These facts are well documented in Walpole, *Memoirs of the Reign of George II*, ii. 35–7.

[18] This and the quotations that follow are taken from *Grenville Papers*, i. 434.

on 13 November. It was not a sudden change of direction but a quiet withdrawal of support from Pitt's extreme positions on policy. He raised no general objection to Pitt's plan for an attack on the subsidiary treaties with Hesse, and latterly Russia, in the opening days of the session. Indeed, the scene was set for a public row with Newcastle on the night of 12 November when Grenville, his brother James, Pitt, Legge, and Charles Townshend failed to attend the eve of session ministerial gathering at the Cockpit.[19] Yet events on the day of the Address proved something of an anticlimax. Grenville played his part in this. He did believe that MPs should question the European alliances as a means of fulfilling Britain's immediate priority of blunting French aggression in North America, but did not consider a wholesale condemnation of ministers and their policies justified. 'We all took our parts in it though upon plans a little different', runs Grenville's account of the debate on 13 November, 'Mr. Pitt attacked the Ministers personally, as well as the measure of the German war: I confined myself to the measure only.'[20] Grenville could not have been disappointed at the outcome of the debate. It lasted until 5.30 a.m. and included many fine speeches from both sides. The government was forced to defend its policies in earnest, as Grenville desired, pinning its faith on the subsidiary alliances on the ground of economy and expediency. Only by gaining the support of European allies, ministerial speakers argued, could a land force capable of challenging France and her ally Prussia be put in the field. Furthermore, they emphasized that the quarrel with France was Britain's and not Hanover's; the part played by the Electorate in the alliances had cost George II much from his own and electoral funds. Grenville answered these points with an excellent speech and a cutting riposte to the basic points of the government case. If this way of dealing with Britain's money and diplomacy could be described by ministers as cautious, he asked the House, 'what would their imprudence be?'[21] This line of attack was directed straight

[19] Walpole, *Memoirs of the Reign of George II*, ii. 47.

[20] *Grenville Papers*, i. 434. This is supported by Gilbert Elliot's account, Minto MSS. 1101, fo. 14.

[21] Walpole, *Memoirs of the Reign of George II*, ii. 52. The quotation that follows is also taken from here.

at the Tory and independent MPs who bore the financial burden of the war through taxation, and Grenville chose good ground upon which to erect his own standard. In his summary of the speeches even Walpole, certainly no admirer of Grenville, allowed that 'Grenville's was very fine and much beyond himself'. Yet it had little effect on the outcome of the debate. Despite an impassioned speech by Pitt to amend the Address, rejecting the policy of subsidiary alliances with Hesse and Russia, the government enjoyed overwhelming majorities in the two divisions that took place. In the first the ministry triumphed by 311 to 105 and in the second, when many MPs had left on seeing the government's strong position, the opposition could raise only 89 votes to 290.[22]

There is no simple explanation of this dismal failure. The government's success was a mixture of luck as well as good management. First and foremost Pitt committed a serious tactical blunder in attacking and then dividing on an amendment to the Address. To attack ministerial conduct of foreign policy in the eighteenth-century Commons always proved a hazardous practice, for MPs were generally inclined to give the government the benefit of the doubt. The depth of support for Newcastle's alliances amongst the independent members of the Commons had also been underestimated. There was disquiet amongst the government's own supporters on the point of subsidizing Russia and Hesse but their protests were much more restrained than Pitt's outburst.[23] His personal attacks on ministers proved unpopular and he would have probably fared better to stick to Grenville's policy of objecting to measures not men. To be fair to Newcastle, he had prepared himself well for the opening of this session of parliament. After his abortive talks with Pitt in September, he tackled the threat of another internal revolt by offering Fox the post of secretary of state and places for the Duke of Bedford and his followers.[24] This ploy of divide and rule worked perfectly in outflanking

[22] *Commons Journals*, xxvii. 298.

[23] See, for example, *Bedford Correspondence*, ii. 170–1.

[24] Fox did not take up the appointment before the opening of the session. This would have entailed his absence from the Commons on the Address, for, in accordance with eighteenth-century parliamentary procedures, MPs had to seek re-election to the House on appointment or a change of office.

the Cobhams. Those recently displaced by Pitt and Grenville at Leicester House found refuge at Court. The wheel turned full circle. On 13 November Fox pinned his colours firmly to Newcastle's mast, receiving round condemnation from Pitt who sarcastically compared the alliance of Fox and Newcastle to the confluence of the two great rivers in France, the Rhone and Saône. All to no avail; the accession of the Bedfords and Fox's promotion gave Newcastle the confidence to proceed with his alliances and rid himself of the Cobhams. On 15 November Fox received the seals of office in the place of Sir Thomas Robinson, to become leader in the Commons by name and rank: on 20 November Pitt, Grenville, and Legge were dismissed and James Grenville resigned his post at the board of trade. To add insult to injury, Lyttelton took Legge's place as chancellor of the exchequer, and the rout was complete.

Pitt was not in the least bit dismayed at the prospect of opposition; indeed he felt more comfortable that his antagonism towards Newcastle could now have free rein. Grenville felt less sanguine about the future, for what chance was there of an early return to office? The blunt truth was that Newcastle had isolated the Cobhams and presently enjoyed a Commons majority of over 200 votes. A long period in the wilderness without the financial rewards of office beckoned Grenville, and nothing occurred before the Christmas recess to dispel this depressing picture. Grenville and Pitt hoped to make an impression at Westminster as soon as possible and decided to campaign on three fronts, on matters directly related to the threat of open war with France in the colonies and Europe. The overall strategy would be to oppose Newcastle's conduct of foreign policy in respect of blunting French aggression on traditional Cobhamite lines. As Pitt put it in early December in a letter to Grenville, 'I think we may, with as much effect, assert our insular plan, by declaring that we mean to enable his Majesty to defend the dominions of England, and not to lay the foundations for continental operations.'[25] This 'insular pain' consisted of three facets: the strengthening of the navy; raising militia regiments to defend Britain, thus

[25] *Grenville Papers*, i. 152.

avoiding reliance on foreign troops; and repudiation of Newcastle's subsidiary alliances for their expense and tacit commitment to military involvement on the continent.

Grenville readily fell in with this plan. It concentrated on policy rather than personalities and contained issues, in particular the expansion of the navy, close to his heart. Furthermore, he believed that it was a sensible way to exploit disillusion with the government's response to French aggression and win over many MPs normally favourable to ministerial policy. The campaign began in earnest on 2 December when Grenville seconded a motion by Lord Pulteney for 'the Encouragement of Seamen and speedily Manning the Navy'. In its wording the motion encapsulated two points: to set the navy on a war footing and by implication, force the ministry into declaring its position with regard to the hostilities in the colonies. In blunt terms, was Britain at war with France or not? In Grenville's view the government was being naïve in not preparing itself for an open conflict which had already begun in North America. The bill was needed, he told the Commons, in 'a fine emphatic speech', for the defence of the realm and to do away with press gangs in the recruitment of sailors.[26] The press, he pointed out, was inhumane and a most inefficient means of guaranteeing naval personnel in times of war. Grenville's time at the admiralty stood him in good stead during the debate for he made the valid point that a much better way of attracting ratings would be to promise a share of all prizes taken to victorious crews. This strong plea for reform and preparedness fell on deaf ears. The government rejected the bill primarily because it jeopardized the peace negotiations with France. It would look bad, one government speaker said, to tell the French Britain's intentions were peaceful while at the same time passing legislation to enlarge the navy. No government spokesman expressed a view one way or another about ministerial attitudes on whether or not war should be declared, and no one in the House seriously expected them to. Revealing diplomatic secrets to the House in the middle of a negotiation could hardly have helped the

[26] Walpole, *Memoirs of the Reign of George II*, ii. 82. The account of the debate that follows is taken from here and that in *Parliamentary History*, xv. 544–616.

cause of peace. The arguments against Grenville's desire to abolish pressing proved less convincing. The simple objection being put was that as this method of manning the navy had not failed the country to date, why change it? In the division that followed the debate the government won the day, for at root everyone wanted the peace negotiations to succeed, and the majority of MPs felt they should not risk harming them with a bill that could be seen as provocative. The majority of 130 votes reflects this trepidation, but Grenville's strong advocacy of enlargement and reform did not go to waste.[27] On 17 May the following year hostilities between Britain and France were officially declared, and on 18 May this same bill passed the Commons without opposition.

Grenville and Pitt did not feel despondent about the outcome of the debate on 2 December. They took heart from the fact that they had won over the debating talents of men like Charles Townshend, James Oswald, and Gilbert Elliot, during the day's proceedings, and that from the size of the minority it could be seen they had also captured many Tory and independent votes. The threat of further opposition, as Walpole noted, certainly forestalled any further ministerial changes. 'As the subsidies are not yet voted', he told Sir Horace Mann on 4 December, 'and as the opposition though weak in numbers, are very strong in speakers, no other places will be given away till Christmas, that the re-elections may be made in the holidays.'[28] Pitt took up the next stage of the campaign therefore in good heart. On 5 December he surprised the House by supporting Lord Barrington's motion for an extra 15,000 troops to be added to the army estimates. There was, however, a sting in the tail. He supported Barrington only on the understanding that an extra 15,000 men was better than no increase at all. In Pitt's opinion 18,000 more troups at least were required with the threat of war hanging over the country.[29] Such a number, he admitted, was more than Britain could afford. But on being asked how this discrepancy in numbers could be bridged, he replied in a well-rehearsed move:

[27] *Commons Journals*, xxvii. 325. The division figures were 211 to 81 votes.
[28] *Letters*, ii. 374 (4 Dec. 1755).
[29] *Commons Journals*, xxvii. 328.

He wished the Government would encourage the Nobility and Gentry to form a militia, as a supplement to the Army. He wanted to call this country out of that enervate state, that twenty thousand men from France could shake it.[30]

Little discussion took place on this idea of a militia in the remainder of the debate, but Pitt achieved his immediate end of airing the idea and preparing the ground for a more formal proposal. This took place on 8 December when George Townshend moved for a Committee of the Whole House to look into the state of the militia laws. Pitt seconded the motion and made a sound case for reforming the existing regulations to produce a trained, equipped, and reasonably paid force of some fifty to sixty thousand men to supplement the regular forces defending Britain. No one attempted to refute the charge that the position over raising the militia was hopelessly confused, and the government saw fit to grant Townshend's request for a committee.[31] This seemed a great success for the opposition. Newcastle gave way initially because he knew that the idea of raising militia regiments appealed to the Tories and independents on all sides of the House whose votes and support were highly valued. For reasons of pride and expense they did not wish to import foreign mercenaries to do a job that they believed native-born Englishmen could accomplish. There was too the added attraction of being seen to act with true patriotism. Great prestige could accrue in the local community by playing a leading role or holding high rank in militia regiments.[32] Also important in explaining the government's retreat on this issue, however, was that Newcastle could avoid accusations of provoking France by proposing that the Committee sit in the new year. Even then Newcastle knew that it would be some time before legislation appeared from the Committee's deliberations. In the event this tactical withdrawal proved justified. It was May 1756, after war had been declared, that a new Militia Bill passed the Commons; then antipathy in the Lords to the clauses setting the figure

[30] Walpole, *Memoirs of the Reign of George II*, ii. 89. This plan was mooted to Grenville in December, *Grenville Papers*, i. 152.

[31] *Commons Journals*, xxvii. 331.

[32] These points are explained in much greater detail by Western, *The English Militia in the Eighteenth Century*, chap. vi.

at fifty to sixty thousand men was still sufficient to baulk the bill's passage that session.[33]

In December Grenville and Pitt did not envisage such problems and delay, and took heart from the simple fact that they appeared to have got their own way on the issue. They were in good spirits to face the final, most important part of the campaign against Newcastle – approval of the Russian and Hesse treaties. These had been laid before the House on 26 November, and 10 December appointed as the day for discussing the content. The opposition came well prepared for this event, for the decision had been taken by several of Grenville's new allies to battle the treaties at every stage of their passage through the Commons. Lord Barrington opened for the government, simply moving to refer the papers and treaties to committee. Thomas Potter for the opposition objected to this on the fundamental grounds that the government had acted in an unconstitutional manner, both in agreeing a treaty with a foreign power and paying subsidies without parliamentary approval.[34] This set the tone for other opposition speakers in the debate, including Grenville. He took up Pitt's argument and 'pointed out the impropriety of referring illegal papers, to see if the Committee would grant money on them; and the impossibility of forming a change in the Committee, instead of giving money, and then considering whether it was legal or not.' He agreed with Potter that the government had violated the Act of Settlement in arranging these treaties and refuted the government's precedents to prove the contrary. In conclusion Grenville put the question that he and Pitt had laboured many times before, could anyone dispute that the treaties 'would engage us in a war for Hanover'? Up to that point the debate had been lively and encouraging for the opposition; in the end however it degenerated into a slanging match between Pitt on the one side and William Murray and Hume Campbell on the other. Pitt began his speech in the style of Potter and Grenville but could not resist for long the chance of launching a personal attack on the characters of several government spokesmen past

[33] *Grenville Papers*, i. 159–61.

[34] The account of this debate and quotations that follow are taken from Walpole, *Memoirs of the Reign of George II*, ii. 105–18.

and present involved in 'avowing Hanover in all'. It could have done little good to the opposition cause. There was disquiet over the treaties but by the end of the debate Pitt's invectives remained the most memorable contribution to the day's proceedings. Such tactics won few friends. In the division the minority increased to 126 but the ministry still enjoyed a majority of 182 votes.[35]

The opposition fared no better in committee on 12 December. The House had rejected the case that the treaties violated the Act of Settlement, forcing Grenville, Pitt, and the other leading spokesmen of the opposition to find fresh grounds for raising objections. In the event they made a praiseworthy effort.[36] Charles Townshend, a recent convert to the opposition, captured the debate with an incisive appraisal of the diplomatic miscalculations behind the treaties. No mention of course was made of illegality; the thrust of the opposition attack now concentrated on the justification for erecting alliances and paying subsidies that provoked Britain's old ally Prussia. True, Townshend declared, France and Prussia were not in alliance, but Prussia had made it clear through its ambassador that it desired peaceful relations with Britain. Did it make sense therefore to plough subsidies into continental alliances, diverting time, money and attention from vital areas of the war in North America and the colonies? This line of argument carried weight and several government spokesmen took care to emphasize that the treaties with Hesse and Russia were defensive. They were designed to meet British commitments to Austria and the Netherlands and protect all British interests in Europe, not just Hanover. Grenville remained unimpressed by this reasoning. He queried the notion of talking about Prussia as it said in the treaty, '*the common enemy*', believing Britain had forgotten the identity of its real antagonist. In a well-researched supporting argument, he then explained that should all the subsidies be paid in full Britain would be £3,180,000 worse off. 'Was our debt reduced only to furnish new subsidies', he chided ministers. 'Why had a mere naval war never been tried?' Though the

[35] *Commons Journals*, xxvii. Grenville acted as a teller for the opposition.

[36] This account of the debate, the division and quotations, are taken from Walpole, *Memoirs of the Reign of George II*, ii. 118–33.

basis of the opposition attack as a whole was much more solid and sensible than that of 10 December, the result proved the same. On the motion for the Committee to agree to both treaties the opposition went down by 289 to 121 votes.

When the Committee reported to the House on 15 December, approval of the treaties became a formality. An exchange between Pitt and Murray preceded the divisions, but no new objections to the alliances were raised. The opposition case rested on the view that continental entanglements should be avoided at all costs and attention focused on the war at sea to secure Britain's colonial possessions. On this the House divided and rejected the opposition case emphatically. The Russian treaty was approved by 263 to 69 votes, and the Hessian by 259 to 72 votes.[37] This represented a depressing show for Grenville and Pitt after the minorities on 10 and 12 December. Solace could be gleaned from the fact that their campaign left a mark on the House. As Patrick Crawford (MP for Ayrshire) told a friend in late December, 'We have had four days' debates upon subsidys, and have had several accidental topicks, much fine speaking, and the utmost exertion of abilitys.'[38] But what of the future now that the ministry had resisted all that Pitt and Grenville could throw at it?

It proved to be a hard slog against Newcastle's large majority in the Commons. In his diary Grenville sums up the remaining weeks of the session thus: 'Upon this principle of resisting the war in Germany, in which we were supported by Leicester House and joined by the Tories, our opposition continued during the remainder of the session.'[39] This by no means portrays the frustration and disappointment that Grenville must have felt in the first half of 1756, as the opposition tried and failed to whip up support for an all-out attack on the ministry's handling of foreign policy. In part this was due to an unforeseen change of diplomatic alliances over the winter, culminating in the Convention of Westminster in January 1756. By this treaty Britain and Prussia guaranteed the neutrality of Germany in the event of war, which indirectly

[37] *Commons Journals*, xxvii. 339.
[38] *Caldwell Papers*, II. (i). 110.
[39] *Grenville Papers*, i. 435.

drove Russia and Austria to seek agreements with France to fulfil their territorial aspirations. Though on paper the odds were not in Britain's favour should war break out, Prussia proved to be a much better ally than either Russia or Austria.[40] Moreover, Prussia offered a better protective barrier for Hanover, releasing Britain from the necessity of constructing European alliances aimed primarily at defending the integrity of the Electorate. In a nutshell this had been the ground occupied by Grenville and Pitt in the autumn of 1755, and though Pitt attacked the Prussian treaty for being an Hanoverian measure, their campaign as a whole sagged over the next six months for want of ammunition.

In the two areas where they sought to embarrass Newcastle in his response to French aggression they received bloody noses. The first occurred on 10 February 1756 when Barrington begged leave of the Commons to introduce a bill to settle pay, rank, and conditions of service for four Swiss battalions raised by an acquaintance of Newcastle and intended to serve in America.[41] Grenville and Pitt attacked the idea of sending mercenaries so late in the day, after the French had done so much damage to British interests in North America. There was little sympathy for this argument among those present. They had committed a tactical error, in fact, for Pitt 'instead of censuring the scheme, dwelt on the tardiness of it, painted the negligence of the Administration'.[42] Grenville then objected to certain words in Barrington's statement, giving the impression that nitpicking was the sole reason for opposing the idea. The majority of MPs clearly believed that the country should be grateful for the help and that it was better late than never. In the division the government won a handsome majority by 165 to 57 votes.[43] Grenville and Pitt learnt nothing from this defeat. They made three further attempts to stop the bill's passage with equally disappointing results. On 20 February they suffered defeat on a technical motion in

[40] Owen, *The Eighteenth Century*, p. 81.
[41] *Commons Journals*, xxvii. 443.
[42] Walpole, *Memoirs of the Reign of George II*, ii. 157.
[43] *Commons Journals*, xxvii. 443. See also Walpole, *Letters*, ii. 395 (12 Feb. 1756).

committee by 213 to 82 votes; on 23 February they failed in a move to have a petition from Massachusetts heard against the Swiss by 158 to 52 votes, and lost the division to stop the third reading of the bill on 26 February by 198 to 64 votes.[44] None of this should have caused surprise. If Grenville had stopped to examine the enthusiastic reaction of the House to the call for Hessian troops to defend Britain earlier that month he would have realized the futility of attacking the Swiss battalions. Military assistance at home and abroad was welcomed with open arms irrespective of its source. The point was underlined on 12 May in the debate on a vote of credit for £1 million to make good the Prussian alliance. The House listened to a well-prepared speech by Grenville on the parlous state of the British armed forces on the eve of a war with France. He cited specific examples of ships unable to sail and inadequate numbers of troops to defend colonial possessions, but in the end he found few converts to his views. Despite a strong element of truth in his accusations nobody appeared to want a change of course in line with Grenville's opinion. No division took place on the sum to be voted, and no one else challenged the ministerial view that this sort of criticism was irrelevant because the time had now come for rallying to the national cause.[45]

The session could have been an unmitigated disaster for Grenville and Pitt had the conduct of the war remained their sole campaign. It has been said that the smaller majorities after the Christmas recess indicate a waning of confidence in the ministry's handling of foreign affairs, but this is hard to justify from the evidence available.[46] The minority votes of spring 1756 also decreased in comparison to the divisions before Christmas, and Walpole seems justified in saying of this campaign, 'the opposition neither increase in numbers or eloquence; the want of the former seems to have dampened the fire of the latter'.[47] What advances were made into the government's majorities arose from attacks on the budget during March. Controversy surrounded one issue in particular, a proposed tax on wrought

[44] *Commons Journals*, xxvii. 463, 467 and 481.
[45] *Lyttelton Memoirs*, ii. 154.
[46] Owen, *The Eighteenth Century*, p. 82.
[47] *Letters*, iii. 399. (23 Feb. 1756)

plate presented to the House on 25 February. Though just one of several new duties proposed in the budget, the plate tax was singled out by the opposition because it affected everyone in the House and offered the best hope of a broad-based attack on the ministry. For two reasons it very nearly succeeded; the manifest ignorance of budgetary matters on the part of the chancellor of the exchequer, Sir George Lyttelton, and the ability of his more knowledgeable opponents, Legge and Grenville, to destroy the economic case for such a tax. When the Committee of Supply reported to the House on 3 March, the danger signals appeared for Newcastle. Legge attacked the tax, demonstrating 'that plate was not luxury, but a national way of hoarding . . . that it would all go abroad, unless the proportions of gold and silver were regulated';[48] and then moved that the resolution covering this tax be recommitted. The House divided and the government scraped home by the skin of its teeth 158 to 156 votes.[49] This represented a remarkable performance by the opposition, and in seeking to explain it Lyttelton complained to his brother that 'Mr. Legge, having declared that he would not oppose it at the first reading, our friends did not attend'.[50] Yet it did not seem to make a great deal of difference whether the government knew of intended opposition or not. On 17 March Grenville 'spoke well' against the tax, pointing out that the weights proposed for possession of plate meant that 'only middling persons were to be rated; the poor and the rich were equally exempted'.[51] On the division over the motion to approve the tax the government again scraped home by the slimmest of majorities, 129 to 120 votes.[52] According to Walpole, Newcastle at this point considered dropping the tax altogether. But, encouraged by Fox, he decided to soften the tax 'till it was scarce worth retaining', and crack the whip over the ministry's absent friends.[53] This action worked perfectly.

[48] Walpole, *Memoirs of the Reign of George II*, ii. 179–80.

[49] *Commons Journals*, xxvii. 494.

[50] *Lyttelton Memoirs*, ii. 508.

[51] Walpole, *Memoirs of the Reign of George II*, ii. 181–2.

[52] *Commons Journals*, xxvii. 530.

[53] Walpole, *Letters*, iii. 408 (25 March 1756). See also Namier, 'The Circular Letters: 'An Eighteenth Century Whip to Members of Parliament', *E.H.R.*, 44 (1929), 588–611.

In the next debate on committing the bill on 22 March the government secured a handsome majority of 245 to 142 votes.[54]

In assessing the value of this campaign to Grenville and Pitt, perhaps all that can be said is that it provided a much-needed boost to morale from an unexpected source. Opposition to the budget frequently cut across the divisions of government and opposition, and, in this case, it seems justified to say that Newcastle suffered no lasting damage to his position from his difficulties over the plate tax. The one storm Newcastle could not weather, however, was the deteriorating strategic position of British forces following the declaration of war with France in May. A groundswell of popular discontent with the ministry's handling of the war dominated the summer months of 1756, especially after news of the loss of Minorca reached England in June. The dismay at this event was universal, and characterized by a comment from one of Grenville's correspondents. 'This morning I heard the whole city of Westminster disturbed by the song of a hundred ballad-singers', he wrote, 'the burthen of which was, "to the block with Newcastle and the yard-arm with Byng".'[55] Politicians and the people at large sought to blame someone for the naval débâcle in the Mediterranean, and the role of scapegoat fell to the hapless Admiral Byng. Grenville remained calmer than most. On hearing the news of the loss of Minorca his immediate reaction was to question whether Byng's squadron did enough to defend the island. But Grenville was too astute to utter a public judgement until he knew the full story. As he told Pitt, 'it may be deemed sufficient to throw the whole blame upon Byng, yet I will venture to say, the other is a question that, in the judgement of every impartial man, now and hereafter, will require a better answer.'[56] Indeed Newcastle did a masterly job in shuffling the culpability for this loss from the ministry on to the admiral by setting up a court martial. Yet it did not prove sufficient to alleviate the anxieties of those members of the government saddled with the job of defending the administration's record in the

[54] *Commons Journals*, xxvii. 538.
[55] *Grenville Papers*, i. 173 (11 Sept. 1756).
[56] *Chatham Correspondence*, i. 163–4.

Commons. The loss of Minorca, and Fort Oswego in Canada coupled with Frederick's abrupt occupation of Bohemia, represented unpromising material for the forthcoming parliamentary session. William Murray became the first casualty. In mid-June the office of chief justice became vacant on the death of Sir Dudley Ryder and Murray insisted on his appointment to the post and elevation to the Lords: 'he meant to rise by his profession, not by the House of Commons', was Walpole's wry comment on Murray's sudden reversion to a legal career.[57] This left Fox to stand alone in the Commons against what would be a furious attack from Grenville, Pitt, and the rest of the opposition. Fox hesitated for several weeks, attempting to persuade Newcastle of the need to strengthen the ministry's hand in the Commons. He failed, and by October Fox had decided that enough was enough and resigned. It proved impossible to go on, Fox told the Duke of Bedford, because the king refused to depart 'from his system of governing by the Duke of Newcastle singly'.[58] Without an able leader in the Commons, Newcastle could see no way of continuing and on 11 November he resigned, recommending to the king that negotiations with Pitt be opened. The old Cobham group was back from the dead.

After surviving some tricky moments at Westminster in the spring, it is perhaps surprising that Newcastle took such a defeatist course of action. Three recent studies of the period ascribe different motives for his resignation. At one extreme it is explained as a simple loss of nerve, for Newcastle had yet to suffer defeat in the House; at the other, standing down is seen as a practical decision based on the knowledge that without a strong leader in the Commons no ministry could have survived the furore caused by the loss of Minorca. Of these, the latter seems the more plausible explanation.[59] Yet it was not simply that Newcastle realized his large majority could evaporate in the wake of British losses overseas. It was the nature and composition of those forces

[57] *Memoirs of the Reign of George II*, ii. 224.

[58] *Bedford Correspondence*, ii. 201.

[59] The three studies are respectively, Browning, *The Duke of Newcastle*, p. 252; Speck, *Stability and Strife*, pp. 264–5, and Clark, *The Dynamics of Change*, pp. 278–82.

lining up to oppose the ministry's policies in the coming session that finally convinced Newcastle to throw in the towel. Isolated groups in opposition like Pitt and the Grenvilles or critical Tories Newcastle could deal with. Over the summer, however, these disparate groups found a common platform on which to air their grievances at Leicester House, the traditional focal point under the Hanoverians of opposition to the Crown. Such a union, if carried into parliament, spelt danger for the ministry because Newcastle's policy of divide and rule would no longer be a certain bet for survival. This situation developed from an unfortunate series of coincidences as far as Newcastle was concerned. Just as British forces lost the initiative and Murray lost heart, Prince George became of age, and required a revised establishment to see him through his adult years until he became king. His grandfather, in a blatant attempt to bring him under his wing, offered him £40,000 per year and apartments at Kensington Palace under his watchful gaze. The allowance was accepted but the apartments refused. The prince had no intention of giving up his independence, and to underline the point demanded that his friend and former tutor, the Earl of Bute, be appointed groom of the stole in his own household. This request went down badly with the king because he detested Bute. He not only saw him as an evil political adviser, cementing alliances with the king's enemies Pitt and Grenville, but George II also believed the court gossip intimating that Bute and the dowager Princess of Wales were lovers. Many within the administration sympathized with the king's view: Bute was, in Waldegrave's opinion, 'an empty shell, with a good person, fine legs, and a theatrical air of the greatest importance'.[60] An unholy row thus developed within the cabinet between Newcastle who sought to placate and give the prince his head, and the king's supporters who resisted Bute's elevation in the prince's household. Not until the first week in October did the ministry resolve the problem in Bute's favour, and by then the damage to Newcastle's position had been done. All summer long Bute and the princess ensured that Leicester House provided an

[60] *Waldegrave Memoirs*, p. 38.

effective vehicle for the opposition on all sides of the House to voice its discontent. It proved a formidable combination, with 'ministers not daring to meet the parliament whilst Leicester House was dissatisfied'.[61] Indeed, with everything else going wrong for Newcastle it proved sufficient to dislodge him from office, for Leicester House would only be satisfied with Pitt in the government.

Herein lay the fatal flaw in Pitt's ill-fated attempt to form a durable administration in the autumn of 1756. Pitt believed that his return to favour stemmed from an acceptance on the king's part that only he of all the politicians at Westminster could save the country. Pitt did not consider the truth of the matter that he had ridden back to the brink of office astride Leicester House and, as Grenville put it, 'intestine dissensions among the principal members of the Administration'.[62] This conceit ran so deep as to push Pitt into extremes when negotiations with the Court opened in late October. Before parliament opened on 2 December, Pitt succeeded in offending most of his friends and foes alike. Not only did he demand a *carte blanche* in forming an administration but also insisted on keeping his ideas about places and policies a closely guarded secret. The king became the first victim of Pitt's insensitive conduct, being forced to accede to demands that two of his favourite politicians Newcastle and Holdernesse be dismissed. When George II then tried to bring Fox and Pitt together, the latter assigned the former such a menial role in the projected ministry that Fox was obliged to resign on principle. In early November Bedford told his wife of the king's anger at such actions, 'I was in with the King this morning, and most graciously received, though I found his Majesty in great wrath, and that I think with reason, for the cavalier treatment he has received from Mr. Pitt.'[63] Worse was to come however, for in the final list of offices Temple took the post of first lord of the admiralty, prompting the king to declare that 'he was so disagreeable a fellow, there was no bearing him, that when he attempted to argue, he was pert, and sometimes insolent; that when he meant to be civil he was

[61] *Waldegrave Memoirs*, p. 67.
[62] *Grenville Papers*, i. 435.
[63] *Bedford Correspondence*, ii. 206.

exceedingly troublesome, and that in the business of his office he was totally ignorant'.[64] It did little for the king's temper that in the few short months that Temple held the admiralty, his actions certainly proved these comments well founded.

Even Leicester House could claim some mistreatment at Pitt's hands. As Pitt prepared to go to his first meeting with the king's representative, Hardwicke, he made it clear to Grenville that he would proceed 'without previous participation with Lord B[ute]'.[65] This line of approach he maintained throughout the negotiations by keeping Bute at arm's length, an ungrateful act in view of the part played by Leicester House in reviving his ministerial career. The bitterest critic of Pitt's actions in late October and November, however, was Grenville himself. Several factors grieved him, and Grenville expressed his enmity without reserve through entries in his wife's diary.[66] Pitt's refusal to consult him over places rankled, as did the fact that every piece of information Grenville heard of the negotiations came from a third party. Most important, however, Grenville felt that he had been passed over yet again in the scramble for offices. Many lists and suggestions for offices passed between Pitt and the future leader of the government, Devonshire, in these weeks, but from the beginning Grenville was marked down for the paymastership. He accepted this designation gracefully, even though Legge's projected return to the exchequer left him uneasy. In early November these plans went awry when it became clear Fox would take no part in the new ministry. A reshuffle occurred therefore, and in the revised plan Grenville found himself with his old office as treasurer of the navy. To add insult to injury, the pay office was divided between two new recruits to the ministry, Potter and Dupplin, whom Grenville considered his inferior by some degree. In Grenville's version of these events no one escaped censure, and it is clear after this

[64] *Waldegrave Memoirs*, p. 95. One modern authority does not dispute Waldegrave's view, see Middleton, 'Pitt, Anson and the Admiralty, 1756-1761', *History*, 55 (1970), 189-98.

[65] *Grenville Papers*, i. 178 (17 Oct. 1756). See also *Border Elliots and the Family of Minto*, pp. 332-3.

[66] The account and quotation that follow are taken from *Grenville Papers*, i. 435-9.

incident that he no longer felt beholden to any of his family or friends, least of all Pitt:

> I have said before that Mr. Pitt never spoke to me again upon this subject, for which his illness at Hayes, where I saw him but twice during this transaction, was the real or pretended excuse.
>
> My two brothers were privy to all that had passed upon this occasion; to them I expressed my surprise and dissatisfaction at a behaviour so contrary not only to the friendship and alliance subsisting between us, but to the engagements of honour and good faith. I cannot say that either of them interested themselves at all in this complaint, or took any other part than to use their utmost endeavours to persuade me to acquiesce to it.

These events promised ill for the new administration. The king's hostility and Pitt's imperious leadership meant that the ministry would stand or fall by its policies, in particular the conduct of the war. On this point Grenville seemed surprisingly well informed. In late October he dined with Richard Glover and 'most frankly revealed their whole plan'. It consisted in three solid facets: 'inquiries into past misconduct, the establishment of a militia . . . and sending back the foreign forces.'[67] What Grenville omitted to say in this conversation was his own feeling about these policies, probably passed on to him through Temple. If this was a deliberate omission, it certainly proved wise to have doubts. Over the next five months none of these policies commanded the whole support of either parliament or the people, and Pitt failed to fulfil the expectations raised by his appointment. The most common reason presented by contemporary observers for this fall from grace was the ministry's reliance on 'the partiality of the Tories' at Westminster.[68] There was a grain of truth in this belief that the administration lacked a broad base of support, but trouble would only occur if Newcastle declared outright opposition to Pitt and the ministry failed to pursue the country-war policy outlined by Grenville in October. For the duration of this short ministry the right balance was struck, Newcastle remained neutral and no embarrassing divisions were forced by discontented allies. Where the administration floundered was in dealing with the practical problems arising partly from

[67] Glover, *Memoirs*, p. 64.

[68] Walpole, *Memoirs of the Reign of George II*, ii. 276.

its policies but also from Pitt's unfortunate illness over the winter.

Pitt was too poorly to attend any of the debates before the Christmas recess, and his absence was keenly felt in cabinet and the Commons. Ministers pressed ahead with the policies outlined by Grenville, but relations with the king went from bad to worse. First a damaging argument between Devonshire and Temple over the Lords' reply to the Address spilt over into a debate in the Upper House. It concerned an addition to the reply that Devonshire supported, thanking the king for sending for electoral troops to defend Britain earlier that year. The cabinet agreed to it as a sensible means of healing the breach in relations between the king and his ministers. Temple was absent from the meeting, however, and much to everyone's dismay spoke against the amendment in the Lords on the opening day of the session.[69] Pitt hoped to avoid such folly in the Commons by dividing its management between Legge and Grenville until he regained fitness to return to the House. But Grenville thought very little of this proposal. Apart from the fact that be believed he could manage the Commons without the help of his old rival, he thought the idea of a division fraught with practical difficulties. Leading the Commons could, in his opinion, only be successfully done single-handed. This Grenville made clear to Pitt when asked to send out the circular letters before the session opened. Grenville considered it was Pitt's own duty to exhort the faithful to attend the House and refused to contemplate taking on the task, even with Legge's help. Grenville did not mind being a spokesman for the ministry but, as far as he was concerned, Pitt could do his own dirty work as leader.[70] If Pitt harboured any doubts about Grenville's disappointment with his position in the new ministry, this letter swept them away.

In two important debates in the Commons before Christmas, on 13 and 22 December, this gulf between Pitt and his lieutenants became all too apparent. The debate on 13 December concerned the plan to order home Hessian troops brought over in the spring. Before the day's proceedings began, Grenville

[69] *Waldegrave Memoirs*, pp. 89–90.
[70] *Chatham Correspondence*, i. 196–8 (18 Nov. 1756).

was obliged to seek clarification from Pitt on what form the intended legislation should take, and hinted that he queried the wisdom of sending back these defensive forces without adequate replacements. In the end he had to accept Pitt's explanation that he 'should not have given into this matter, if I had not seen the ground clear'.[71] But when Barrington moved the bill for preparing quarters for the Hessians until their departure, Grenville's worst fears were fulfilled. Lord George Sackville and Henry Seymour Conway, both military men, questioned what would be put in their place, and, more to the point, would any replacements be immediately operative? Grenville of course knew that they would not, answering that the defence of the realm would be entrusted to the militia, and a bill would be moved the following day for raising the 65,000 men required.[72] This rather lame reply, according to Henry Digby (MP for Ludgershall), 'was applauded by the Tories and Tories only'.[73] Grenville felt most uneasy at this turn of events. It is clear that he told Pitt of trouble ahead but only received a note of encouragement in reply.[74] A week elapsed before the storm. On 22 December the army estimates came before the House and much to everyone's surprise they covered payment for the troops destined to return to Germany. To correct this anomaly Legge told MPs that these troops should not however 'be looked upon as part of the permanent strength of this country'.[75] This slip of the tongue opened the door for Sackville to ask why the House had therefore voted for 49,000 when 'we had not, without the foreigners above 36,000, and that when all the necessary garrisons were supplied we should not have above 18,000 to act in case of invasion, and that all the force we could send to America must be deducted from that 18,000.' In one fell swoop he had undermined the whole ministerial case for sending the electoral and Hessian troops home, and subsequent speakers demanded evidence of military preparedness. Grenville, now

[71] *Grenville Papers*, i. 187 (12 Dec. 1756).

[72] George Townshend again moved the bill on 14 December, *Parliamentary History*, xv. 782.

[73] H.M.C. 8th report, p. 223.

[74] *Grenville Papers*, i. 188 (15 Dec. 1756).

[75] The account and quotations that follow are taken from Henry Digby's report in H.M.C. 8th report, p. 223.

on the spot, replied that he could not reveal such confidential information as it would give succour to the enemy. But he did promise the House that the foreign troops would not be sent back secretly, though no formal declaration would be made in the House. With this unsatisfactory explanation the debate closed, for Grenville had no idea how Pitt intended to handle the timing of these measures. Without its leader, the ministry could do nothing but stumble along from one difficult situation to another. After these debates it soon became clear that a vast number in both Houses opposed such a large militia force, preferring a reduction to something like 32,000 men, and, moreover, saw it as their duty to promote petitions and addresses demanding the retention of foreign troops as defence against invasion. Digby summed up the ministry's predicament well as it retired for the Christmas recess: 'It seems by the appearance of the House that this new Administration has the Tories and nothing but the Tories to support them; and I do not think they seem extremely zealous just now.'

In one sense Digby's remark appeared correct: the Tories as a body were disappointed with the practical results of these policies. After the Christmas recess ministerial policies proved no more successful than before the holidays. The opponents of the Militia Bill finally managed to whittle it down to half the number demanded by the ministry; the budget proved inadequate and Pitt found himself unable to promote the grandiose schemes for prosecuting the war he had canvassed in opposition. The government dispatched a small force to North America in early March but the wide-sweeping naval strategy close to the ministry's heart remained a pipe-dream. Worst of all, the inquiry into past misconducts, of which Grenville told Glover the previous October, went disastrously wrong. The misconduct concerned, of course, was the loss of Minorca. In favouring an inquiry the ministry hoped to show that Newcastle and his ministers should shoulder the blame for this loss and not the navy and its commanders. For this Pitt, Grenville, and Temple have been attacked unmercilessly. Waldegrave wrote in his memoirs that 'the popular cry was violent against the admiral, but Pitt and Lord Temple were desirous to save him: partly to please Leicester House, and partly because making him less criminal, would throw greater blame on the late administration.'[76] More recently one historian described the inquiry into the loss of Minorca as 'absurd and utterly hypocritical'.[77] In the way that the ministry presented its case there is no doubt that revenge on Newcastle appeared the sole motive. It became imperative before laying any papers before the House to obtain a reprieve for Byng and establish his innocence. On 27 January 1757 a court martial found Byng guilty of a capital offence and sentenced him to death. Endeavours were then made through personal appeals and official addresses to persuade the king into granting a reprieve. George II proved unsympathetic however and in a message read to the Commons on 25 February rejected the appeals for clemency.[78] This brought an immediate response from Byng's supporters who, immediately after the king's message had been read, moved that those officers involved in the court martial be released from their oaths of secrecy. They felt that not only could the legal case for condemning Byng be undermined but that this was Byng's last chance for escaping the firing-squad. The motion was in fact approval with cabinet support on 28 February by 153 to 23 votes, but it failed to save the admiral.[79] Those involved in the court martial refused to divulge any relevant new facts and the whole proceeding only succeeded in offending the king and leaving the admiral's supporters, like Grenville, open to attacks of opportunism in using the hapless Byng to discredit their enemies.[80]

In the ministry's defence, it must be pointed out however that these were tactical errors; to say that Grenville, Pitt, and Temple did not particularly care about Byng's fate is unjustified. After careful study, Grenville saw no blame attaching to Byng's behaviour. During an exchange of letters with his naval friend George Rodney during 1757, they both concluded that the forces under his command were quite inadequate for the job demanded by the admiralty.[81] This

[76] *Waldegrave Memoirs*, p. 91.

[77] Langford, 'William Pitt and Public Opinion, 1757' *E.H.R.* 88 (1973), 69.

[78] *Parliamentary History*, xv. 803–4.

[79] *Commons Journals*, xxvii. 740.

[80] For further background on this see Peters, *Pitt and Popularity*, pp. 69–71.

[81] See, for example, Rodney's letter of 21 October 1757, *Grenville Papers*, i. 221.

is the judgement that Byng's enemy, Walpole, also reached in a long and extremely enlightening examination of the case.[82] Yet in the end no amount of political manoeuvring could save Byng as the inquiry into past misconducts, the hostility of the king, and the understandable procrastination of Newcastle's followers in these proceedings doomed them to failure. Paradoxically, despite this dismal performance in Byng's case and the general mismanagement of domestic measures, the government came under very little pressure at Westminster in these months or suffered a critical loss of popularity in the country at large. Paul Langford offers a good explanation of this anomaly by emphasizing that achievements came secondary to what the government's policies represented in these times of war, 'it was the record of a truly patriotic ministry, the unmistakable work of a country administration'.[83] Naturally the king did not see it this way. He equated the ministry's poor performance with lack of support and popularity, and by April only needed a ready made excuse to rid himself of his hated ministers. The king found it in April when the Duke of Cumberland refused to take command of British troops in Europe while Pitt remained in the ministry. Asked to choose between the two, George II needed no prompting to sack Pitt and replace Temple at the admiralty with Winchelsea. In the wake of these changes Cumberland left for the Continent a happy man, probably expecting his ally Fox to be in power when he returned. It proved a vain hope. Pitt's dismissal produced a feeling of outrage, in and out of parliament, strong enough to prevent George II and Fox from forming an alternative ministry.

Grenville knew little in detail of these events in March or April. He relied for his information as usual on Temple, who told him in his best arrogant style on 8 April not to rush to London 'for your own resignation, it may keep cold till you return'.[84] Grenville's role in the administration had assumed importance only from his being asked to deputize for Pitt in the Commons. It was an odd twist of fate therefore that in the weeks of confusion from April to July when

[82] *Memoirs of the Reign of George II*, ii. 284–301.
[83] 'William Pitt and Public Opinion', p. 69.
[84] *Grenville Papers*, i. 193.

no one could be found to lead the government, Grenville should be most active in parliament. On 19 April the committee inquiring into past misconduct began to consider papers relative to the loss of Minorca. All the documents talked of and laid before the House since the turn of the year were to be read to members. At their preparation Grenville, according to Glover, 'seemed to have taken some pains', for he wanted the blame to fall on the ministers of the previous government, especially Fox.[85] It took over a week to complete this monotonous task, the climax being reached on 3 May. Three divisions took place on this day, one on a motion by the Court to show that 'no greater number of ships of war could have been sent considering the state of the navy'; two more on motions by George Townshend to demonstrate that the government sent inadequate forces to defend Minorca. The government now leaderless in the Commons won the first division by a majority of 78 and the others by 212 to 134 and 210 to 127 votes. All the effort Grenville expended on this inquiry had been entirely wasted; when it reached its conclusion 'the committee was so tired that few spoke upon the main question when we came to it'.[86]

A similar fate befell Grenville's other legislative project in this short spring session, a bill to secure regular payment of seamen's wages. No one doubted his motives on this point. The bill, based 'on a good-natured principle' aimed to eradicate one of the crippling grievances in seamens' conditions of service. The traditional method of payment entailed keeping back a portion of the wage for supplies on board and tying seamen to their ports of disembarkation by irregular settlements of their tickets' of service entitling them to pay. All sorts of abuses and professional rackets had grown up around these seamens' tickets, none of which helped to attract much needed manpower. Grenville also hoped that full and regular pay would help recruit sailors without recourse to the press, and enable serving men's families to receive better support than hitherto. The bill, like Pulteney's attempted reform before it, met some

[85] Glover, *Memoirs*, p. 91.

[86] This account and the division figures are taken from Yorke, *Hardwicke*, ii. 358–9.

entrenched opposition, but in this case it did pass the Commons by 60 to 42 votes.[87] When the bill reached the Lords however, Lord Winchelsea, the new first lord of the admiralty, opposed reform outright. His grounds for complaint rested on personal and political animosity to Grenville and his family. To his credit Bedford, then no ally of Grenville, spoke of the bill's merits in defiance of factional interests and against his friends. But in the end Winchelsea had his way: the bill went to defeat by 23 to 18 votes.

This miserable end to the short parliamentary session of the spring turned Grenville's mind against Westminster for the next six months. He deliberately retired to Wotton until the opening of the new session in November, sulking about his raw deal over the navy bill. Grenville played no part in the negotiations for a new ministry during May and June, except to waive his pretensions to the office of chancellor of the exchequer he had long coveted.[88] There appear to be two explanations for this unexpected gesture: first, Grenville viewed Pitt's attempt to claim the exchequer on his behalf as a cosmetic gesture in order to make up for earlier disappointments; second, Grenville simply no longer cared about promotion through Pitt's endeavours. He had suffered Pitt's erratic and disdainful attentions for so long that scepticism overcame Grenville's good political judgement. In either case Grenville felt wronged and no amount of cajolery by Pitt and Temple could bring him out of this dudgeon. From the evidence available Grenville seems to have had little cause for doubting Pitt's motives. In May the talks between Pitt, Newcastle, and the king's representatives foundered on this claim of the exchequer for Grenville, and only proceeded when Grenville waived his pretensions to this office. Furthermore, not involving himself directly in the negotiations only compounded the mistaken impressions Grenville gleaned from correspondence. When the new administration took office on 29 June, with Pitt and Newcastle at its head, Grenville simply returned to his old post as treasurer of the navy. He had again hoped for something better from the reports sent down from London,

[87] This and the division that follows appear in Walpole, *Memoirs of the Reign of George II*, iii. 19–20.

[88] Ibid., p. 15.

but his conduct made it likely that he would be staying put. At that stage in Grenville's career this can only be seen as a severe set-back, for he still lacked cabinet rank and would only be involved on the periphery of decisions over policy. Such a situation was guaranteed to increase Grenville's frustration and anger over his thwarted political ambitions. Over the next three years he had cause to regret his flight in this time of political turmoil, and by 1760 was forced to rely on the death of George II to extricate himself from the result of this misjudgement.

There is no doubt the Pitt–Newcastle administration promised a great deal on its formation, but it was no more than that. Pitt's attempt to run an administration single-handed had been an ignominious failure and as soon as war broke out in 1756 Newcastle's government too had floundered. The country was fortunate that by chance this shotgun marriage worked well. Pitt's mighty schemes and bellicose posturing proved invaluable in the dark days of the war against France over the next three years, while Newcastle was left to organize the Commons and undertake the not inconsiderable task of raising the vast sums of money needed to support Pitt's global battlefront.[89] Where did Grenville fit into this scheme of things until its demise in October 1761? In the main he remained an outsider; after a brief flurry of parliamentary activity over the winter of 1757–8, he concentrated his efforts on naval matters. Keeping a low profile brought its rewards initially. Much to his delight Grenville managed to push through the bill reforming payment of seamens' wages rejected in the previous session. The same legislation was laid before the House on 8 December 1757 and met no opposition. On 24 January 1758 Grenville read the bill to the House and made an impassioned plea for ending the abuses arising from the irregular payment of seamens' wages by the ticket system. In a moving passage of his speech, he portrayed the practical and financial difficulties of a seaman's family when the breadwinner spent long periods away without any means of remitting part of his wage home:

[89] Speck gives a succinct account of Pitt and Newcastle's working relationship from 1757 to 1760 in *Stability and Strife*, chap. 12.

I am sorry to observe, Sir, that the derision with which our brave sailors are treated, and the slight esteem in which their profession is held, in proportion to its importance, are no small aggravations of the severe hardships which they so unjustly experience in themselves and their families.[90]

This strong case proved sufficient to head off any opposition in the Commons, but Grenville had to attend the second reading in the Lords, and give evidence, before they allowed its passage in the spring. It became law in April that year. Overall Grenville had good cause to feel satisfied with this reform, though judgements do vary as to its practical effect. The naval historian, Baugh, says it was a worthy bill but did little in practice to improve recruiting outside the press; the simple reason for this being that in wartime sailors as merchantmen received between 50 to 60 shillings per month as opposed to the naval rate of 24 shillings per month.[91] Yet this was not the sole criteria for moving the bill, and Grenville would certainly have been pleased that his contemporaries saw its real value in other terms. John Almon, writing some fifteen years after its passage, wrote:

How defective and imperfect soever this Act may appear to those who have accurately and maturely considered the subject, it certainly flowed from a principle of humanity; and the salutary effects of it soon appeared in considerable sums of money, which even the common seamen, remitted to their poor families, in different parts of Great Britain and Ireland.[92]

From this point on there is little to say of Grenville's political career until George II's death in October 1760. His public rôle as Pitt's deputy in times of illness and absence remained unchanged for the next three years. Grenville rarely spoke in the House after the spring of 1758, and, in truth, there was little need for him to do so. The Commons, as Walpole observed, had only the defeat of the French in its sights: 'Now the Parliament is met, you will expect some new news; you will be disappointed: no battles are fought in Parliament now – the House of Commons is a mere war-office, and only sits for the despatch of military business.'[93]

[90] *Parliamentary History*, xv. 840.

[91] *Naval Administration in the Age of Walpole*, pp. 228–9.

[92] *Debates*, v. 222. Grenville's evidence to the Lords had in fact concentrated on these abuses, *Lord Journals*, xxix. 267.

[93] *Letters*, iv. 320 (16 Nov. 1759).

His last solid contribution in the spring session of 1758 was to support a move by the ministry for a bill 'for giving a more speedy remedy to the subject upon the writ of Habeas Corpus'. It passed the Commons without trouble on 17 March but met opposition when carried to the Lords in June.[94] After a great deal of debate on the wording of the bill a compromise was reached, approving a suggestion by Hardwicke that a drafted bill be presented the following session. This proved a successful ploy by the bill's opponents because it was 1816 before this amended draft saw the light of day![95] Pitt did consult Grenville over the Address in November, but after this, until parliament met in 1760, there is a void in Grenville's personal and political life. In the winter of 1758 his eldest son Richard, aged seven, fell ill and after a long and agonizing illness he died on 7 July 1759. During these months Grenville paid little attention to parliamentary affairs; his colleagues sent regular reports of debates and several congratulatory letters for naval victories which Grenville helped plan and equip. They evinced little response but a polite and bland reply. Grenville refused to leave his son's side where it was possible to do the routine work of his office in the admiralty. As the country responded rapturously to the series of victories gained by British forces in 1759, Grenville could not have had less interest in events at Westminster.

The significance of these years lies not in Grenville's public utterances but in his private misgivings about Pitt's conduct of the war and behaviour towards colleagues. After the disappointments he suffered in the autumn of 1756, Grenville already had a generous store of antipathy towards his leader, and nothing occurred before George II's death to heal these wounds. On the contrary the gulf between Grenville and Pitt widened considerably in the last two years of the old king's reign. On two fundamental points their views differed: Pitt's cavalier attitude to the enormous sums of money being spent on the war, and relations with Leicester House. Parsimony was by necessity a way of life in the Grenville household and he deliberately carried this

[94] *Commons Journals*, xxviii. 159 and Yorke, *Hardwicke*, iii. 5.
[95] Walpole, *Memoirs of the Reign of George III*, iii. 120–1.

approach to financial affairs into office. No exception was made for the demands of the war against France. As early as August 1757, Pitt and Grenville exchanged letters over the huge increase of public money needed to prosecute the war in the coming year. In his first reply on this matter the reasoning behind Grenville's statements on the conflict appear clear and to the point. Increased expenditure of the magnitude envisaged by Pitt could be justified on condition that each military commitment was 'thought fully weighed and considered'.[96] But he did feel Britain's strategic intentions should be defined and adhered to. Pitt accepted this advice in good faith, yet by October he had gone well beyond Grenville's position. In his view conquest could not be gained cheaply. When questioned about Britain's mushrooming commitments around the world, Pitt told Grenville that 'immense expense is unavoidable, and the heavier load of national dishonour threatens to sink us with double weight of misfortune'.[97] As the sums asked of parliament grew larger with each subsequent engagement, Grenville's disillusionment with Pitt's policies deepened accordingly. By December 1759 this lack of faith had turned into active discussions with other colleagues about peace, and the terms Britain must demand in any negotiation. There was no chance of compromise once Grenville crossed this bridge, for as Pitt wrote to him on 18 October 1760 with regard to an Address from the City 'such generous and warm assurances of supporting the war, cannot but give the highest satisfaction to Government'.[98]

At first the problem over Leicester House was a separate issue. After the Pitt–Newcastle administration had been in office several months, Prince George and his mentor Bute began to complain of Pitt's high-handed behaviour towards Leicester House. Their initial displeasure concerned the fact that Pitt would not consult them over policy decisions. No agreement existed that he should; the prince and Bute were of the opinion however, that as they had played such a large part in rescuing the political careers of the old Cobham group in 1756, it was the least he could do. The

[96] *Chatham Correspondence*, i. 245.
[97] *Grenville Papers*, i. 227.
[98] Ibid., p. 355.

prince did not even demand that their views be heeded: 'I suppose you agree with me in thinking that as Mr. P[itt] does not now choose to communicate what is intended to be done', he complained to Bute in December 1758, 'but defers it till executed, he might save himself the trouble of sending at all.'[99] This tone soon changed; in the following summer Pitt and Leicester House fell out over the treatment of Lord George Sackville, concerning his role in the battle of Minden. From hearing the news of the encounter Pitt believed that Sackville was guilty of cowardice. On his return to England for a court martial Pitt refused to speak to him, and fully supported the king's decision to forbid him attendance at Court or hold any civil or military post in the future. Bute and the Prince took an opposite line, providing support for Sackville when, as a result of the king's actions, he became a social outcast. It was during this incident that Grenville drew closer to Leicester House. Unlike his brothers, Temple and James, or Pitt, Grenville did not believe that Sackville committed a cowardly act. He did not share Bute's friendship with Sackville, but this did not deter him from keeping an open mind about the issue and offering the same sort of sympathetic support provided by the prince.

Indeed, it is a common but mistaken assumption that as Bute and the prince grew disillusioned with Pitt, Grenville stood loyal by his brother-in-law.[100] In reality Grenville was drawn further into the confidence of Leicester House in late 1759 and early 1760 because Bute and the prince came round to the view that Pitt's endless expense in fighting the French was having a disastrous effect on the country.[101] The extent of Grenville's rôle in directing them down this path is uncertain, but there is no doubt that he had many opportunities to encourage such beliefs. Grenville had enjoyed good relations with Bute since his dismissal from the ministry in November 1755. Bute had wrote then, 'I must tell you my worthy friend . . . 'tis glorious to suffer in such a cause and with such companions.'[102] This

[99] *Letters from George III to Bute*, p. 18.
[100] This is done in Mckelvey, *George III and Lord Bute*, chap. vii.
[101] See, for example, the letter from the Prince of Wales to Bute on 5 Oct. 1760, *Letters from George III to Bute*, p. 18.
[102] *Grenville Papers*, i. 148–9.

friendship developed in the last years of George II's life, and can be traced through Grenville's contacts with other members of Leicester House, like Sir Henry Erskine, who provided him with political snippets from the prince's household.[103] This bond with Leicester House was sealed when Grenville brought his friend Charles Jenkinson into Bute's orbit in 1756. Jenkinson was an able administrator, looking for promotion from the beginning of this association. After serving with Lords Harcourt and Holdernesse, he realized his ambitions in March 1761 when he became Bute's under-secretary of state at the northern department.[104] A look at the correspondence of Grenville and Jenkinson in these years not only reveals a strong friendship and political association but also deep-rooted antipathy to Pitt's policies. Their misgivings over Pitt s handling of the war were openly stated. In September 1758 Jenkinson remarked that mismanagement of the country's strategic interests was always a symptom of 'unsteady Administrations'.[105] Grenville did not challenge this, as his hopes for the future were pinned firmly to the reversionary interest. When George II died, Grenville had become identified with the political opinions of Bute and the prince. It was not simply a matter of agreeing on the need to think about peace and putting an end to an expensive war but also the fact that, unlike Pitt and Temple, Grenville never forgot the debt he owed Bute and the Prince of Wales.

At the close of George II's reign it appeared that Grenville had at last found a means of freeing himself from Pitt's malignant influence on his political fortunes. In the six years since Pelham's death Grenville had experienced the extremes of high expectation and bitter disappointment in the search for preferment. In part Grenville must be held responsible for his misery in these years. He succumbed too readily to the arrangements made for him at the time of Pelham's death, and again when Newcastle resigned in the autumn of 1756. If the pent-up anger accumulated after these events had been turned into positive action and protest during the changes, he might have achieved the senior posts that he

[103] *Grenville Papers*, i. 189–91.
[104] Jenkinson's early career is described by Jucker, *Jenkinson Papers*, intro.
[105] *Grenville Papers*, i. 272.

craved. Being reasonable and modest, showing due deference to the wishes of Pitt and Temple cost Grenville dear. For solace, Grenville could, however, weigh these disappointments against solid achievements on other fronts. The post of treasurer of the navy may not have carried cabinet rank but it proved the financial salvation of Grenville's large family, and brought him the valued prize of economic independence from his brother. Moreover, in these years Grenville established himself as a formidable speaker and debater in the House of Commons, fulfilling a prophesy made by Richard Glover seventeen years earlier. In December 1743 Glover met Grenville for the first time and wrote, 'I waited on Lord Chesterfield at dinner on the day appointed, when I met Pitt, Lyttelton and George Grenville, who I believe will make the most useful and able parliament man of the three.'[106] His knowledge of Commons' procedure and blunt commonsensical approach to legislation drew the respect of his contemporaries, friend and foe alike. The common assumption that Bute somehow picked Grenville from obscurity and launched his ill-fated parliamentary career of the 1760s could not be further from the truth. If he could rid himself of the faint heart that baulked his ambitions in the past, Grenville had everything to gain from the new reign.

[106] *Memoirs*, p. 16.

Chapter IV

A New Dawn, 1760–1763

Grenville received the news of George II's death at Wotton in late October 1760, and made immediate preparations to travel to London for the royal funeral on 11 November. During the ceremony Grenville marched and conversed with Horace Walpole, and probably learned for the first time that changes were in the air. At this early stage they concerned measures rather than men. The young king, George III, was not so headstrong as to cast Pitt and Newcastle aside at the height of their success in the campaign against the French. His pace would be slow and deliberate. The battlelines were drawn on 27 October at the first meeting of the privy council in the new reign. George III had drawn up his address for this small gathering without consulting Newcastle or Pitt, but when the latter saw the draft he insisted on crucial changes in the wording of the speech. The king originally intended to say that he came to the throne 'in the midst of a *bloody and expensive war*' to which he hoped to bring 'an honourable and lasting peace'. With a mixture of bullying and persuasion that the king neither forgave nor forgot, Pitt insisted that *bloody and expensive* be changed to '*just and necessary*' and the words '*in concert with my allies*' be added to the word's 'lasting peace'.[1] Grenville could not have been disappointed on hearing of this indication of the king's views. He had felt for some time that the cost of war in lives and money was too great for Britain to bear. Indeed, the conduct of the war and struggle for peace would dominate national politics over the next two years, and it proved to be a dividing line not only between George III and his ministers but also between George Grenville and his family's political connections.

[1] *Leicester House Politics*, p. 215. This was also Lord Bute's first attendance at a privy council meeting, having just been sworn in.

The first few weeks of the new reign passed with leading politicians in a state of high expectation and uncertainty: as Lord Egmont wrote on 16 November,

> People most puzzled what to think . . . The king is advised to keep every body at a distance but with show of great civility and courtesy but to endeavour to fix a character of being immoveable in his determinations, and that Bute should be the sole favourite and director, and Bute and the P[rince] really seem to direct all his thoughts and actions.[2]

Before Sir Lewis Namier and John Brooke exploded the myth of George III attempting to establish a new constitutional order in the first years of his reign, such comments formed the basis of a plausible conspiracy theory against the rights of parliament. Yet in these early days of the reign Grenville expressed no fears about Bute's powers behind the throne. He shared the same fears and expectations as Egmont, and his personal record of national political life over the next two and a half years displays a similar practical concern over whether seeds sown during the minority would bloom now the new king was on the throne. The ousting of George II's leading ministers and their replacement with the young king's lieutenants was a natural progression from the opposition cabals at Leicester House. Grenville attached himself to Bute in the old reign because he expected rewards in the new. Loyalty and patience were all that he required before these dreams became reality. After the privy council on 27 October, Newcastle and Pitt had audiences of the king and agreed to continue in office. The arrangement stood, however, without any firm or lasting declaration of support from the king, and soon proved unsatisfactory. After several indecisive weeks of policy-making it became clear that the king favoured the recommendations of peace being urged by Bute over the more aggressive stand of Newcastle and Pitt. It was not long before this division of opinion became known at Westminster, producing an intriguing situation for those ambitious politicians determined to advance in the new order. Two questions influenced their choice of allegiance. On the one hand, nobody could deny Pitt and Newcastle the honour and popularity of delivering the country from

[2] *Leicester House Politics*, pp. 226–7.

the jaws of defeat in the dark days of the war. Was it not just therefore that they should be allowed to conclude the war and guide terms for the peace? On the other, what could be gained from supporting their policies when the king obviously wanted the peace advocated by Bute? Running against the tide of royal opinion usually spelt disaster, and once George III had decided on peace it became only a matter of time before he would have his own way.

Grenville wasted no time in pinning his colours to the royal mast. At Pitt's behest he readily agreed to draw up the Address for the opening of parliament on 18 November 1760. It proved successful; with the only exchange of note taking place between Pitt and his friend Beckford, when the latter charged that the war effort had fallen into a 'languid' state.[3] Grenville wished to go further than this token effort, and later that month during the Committee on Ways and Means opposed an administration measure levying an extra 3*s*. duty on a barrel of beer.[4] There is no doubt that he consulted Bute before this move, using Charles Jenkinson as the intermediary.[5] Their motives were mixed but coincidental. For Bute the tax provided an opportunity to attack Legge, the chancellor of the exchequer. The two had fallen out over election matters and Bute was actively seeking the means to remove Legge. To Grenville, opposition to the tax was a simple declaration of faith in Bute and the young king, without compromising his own principles. 'When it was mentioned to Grenville', wrote a friend two years after the event, 'he disapproved it, and he of all men I conversed with alone disapproved it.'[6] In addition, it must also be said that Grenville was not averse to undermining Legge's position. For over a decade Legge had gained ascendancy in promotions over Grenville, and, should he be dismissed, the latter would be eminently qualified to take a place that Pitt once promised but never secured.

This represented a desperate throw for a politician approaching his fiftieth year. It brought the disapprobation of '*all* his brothers' upon him, and should Bute behave like

[3] Walpole, *Letters*, v. 4.
[4] Walpole, *Memoirs of George III*, i. 26.
[5] *Grenville Papers*, i. 356 and Wiggin, *Faction of Cousins*, p. 235.
[6] *Jenkinson Papers*, p. 141.

Pitt, Grenville could find himself without any prospects of preferment at all.[7] This is not the picture of a politician who, in Sir Lewis Namier's view, had by 1760 become tired of the political game and wanted to retire to the neutrality of the speaker's chair in the House of Commons.[8] There is no record in 1760 of Grenville ever expressing the wish to fill this post, which would become vacant on Arthur Onslow's retirement at the general election of 1761. A more plausible explanation of Grenville's conduct in the winter of 1760–1 is offered by John Almon who wrote about these events in his memoir of Pitt. In this account, the overtures to Newcastle about the speaker's chair in January 1761 were a means of hedging his bets for the future. As the chronology of events shows, the principal target was Legge's post of chancellor of the exchequer, and Grenville only turned all his attention to the speaker's chair when his ambitions in that direction were thwarted. This happened in March 1761 when Bute rid himself of Legge but could not deliver the exchequer to Grenville. In a compromise with Newcastle and Pitt, Lord Barrington took Legge's place, and Bute himself became secretary of the northern department. Grenville did not blame Bute for this further disappointment: 'When it was know that Mr. Legge was to be turned out, Mr. Grenville expressed to his brothers his desire to succeed Mr. Legge; but Mr. Pitt took no notice of his wishes; upon which a coolness commenced between them.'[9]

As it happened Grenville failed to land either of these prizes; though, unlike Pitt, Bute did not leave him high and dry. Grenville received a considerable reward for his show of loyalty. In February 1761 he became a member of the 'nominal' cabinet. This represented an act of faith on Bute's part, for although it did not give Grenville access to all cabinet documents in circulation it did represent a privilege usually denied junior office-holders. His situation improved further in March 1761 when Grenville's political ally Charles Jenkinson became Bute's private secretary with instructions to communicate privately to Grenville 'all

[7] Cited in Wiggin, *Faction of Cousins*, p. 235.
[8] *House of Commons*, ii. 538.
[9] Almon, *Anecdotes of the Life of William Pitt*, i. 304.

things of importance'.[10] This did not entirely satisfy Grenville, and for some weeks to come he insisted in consultations with Bute that his destiny lay in becoming speaker of the House. It took a great deal of work by Bute and the king to persuade him otherwise. They were sincere in keeping him back, Jenkinson wrote on 24 March, for an office 'where your abilities and influence might be so serviceable to the king and the Public'. In view of his past experiences it is easy to understand why Grenville found such assurances difficult to swallow. Grenville's confidence was bolstered initially by his wife, 'you will have the comfort of being soon at the foundation of knowledge' she wrote on 6 March in praise of Bute's actions.[11] Grenville himself mellowed once the detailed news of government business began to flow from Jenkinson.

By the summer relations between Grenville and Bute could not have been better. Grenville was in a position by this stage to recommend friends for posts in the royal household.[12] They also saw eye to eye on policy. Bute's efforts at opening peace negotiations with France at the turn of the year had the full backing of Grenville. They both opposed Pitt's expensive policy of unlimited aggression to defeat France in every part of the globe. This did not imply that Grenville supported the peace-at-any-price opinions of Newcastle. On the contrary Grenville's discussions with Jenkinson over Hans Stanley's peace mission to Paris mark him out as a very tough negotiator. His principal objection centred on Newcastle's pet schemes of expensive subsidies to European powers, especially those to Frederick the Great in Prussia. Though he advocated peace and an end to such subsidies, however, Grenville did not want to abandon what he saw as Britain's true strategic aims. He supported the successful expeditionary force that captured Bellisle in the spring of 1761, and made it clear to Jenkinson that ending subsidies in the cause of peace should not be construed as a sign of weakness. Valuable prizes taken from the French were not to be sacrificed at the conference tables. During

[10] This and the quote that follows are taken from *Grenville Papers*, i. 361 and 357.

[11] B.L. Add. MSS 57804, fo. 25.

[12] Grenville to Bute, 19 July 1761, Bute MSS. 3/23/1.

May and June 1761 a controversy raged in the press over whether or not Canada should be given to the French in return for keeping possession of the rich sugar island, Guadeloupe.[13] In Grenville's view such talk was nonsense. As Britain held both Canada and 'that very material article of Guadeloupe', why make concessions when lesser possessions could be traded with the defeated enemy?[14] There is a great deal of similarity in Grenville's attitude on these issues to opinions expressed in a popular contemporary pamphlet opposing the continuation of the war entitled, *Considerations on the Present German War*. The author, Israel Mauduit, argued the same case as Grenville: an end to the continental campaign and a firm hand in the peace negotiations now Britain's aims in the war had been fulfilled. The tract sold in great numbers and, as Egmont put it, gained credence 'in a very surprising manner'.[15] From this it did appear that Grenville had made a sound political decision. The country was tiring of the war, which enabled Bute and his followers, like Grenville, to exploit the groundwell of popular feeling in favour of peace.

Grenville retired to Wotton in May after the dissolution of parliament in buoyant mood, spending the early summer overseeing his return for Buckingham at the general election and reading over Jenkinson's regular reports from London. By the end of July, when Grenville returned to London, his mood had darkened considerably. The peace initiative did not produce the desired early settlement. On 25 July Jenkinson wrote to Grenville that negotiations had reached an impasse, compelling the British government to issue an 'ultimatum' to the French in an attempt to bring matters to a head.[16] Problems existed on all fronts, with disputes over fisheries off Newfoundland, fortifications in French ports, and possessions in India. The French received the ultimatum coldly and responded with one of their own 'drawn in very strong terms', resisting all moderation on the

[13] The background to this dispute is explained in Rashed, *The Peace of Paris*, pp. 68–70.

[14] Grenville to Jenkinson, 28 June 1761, *Jenkinson Papers*, p. 10.

[15] *Leicester House Politics*, p. 227. Other popular publications against the war are documented in Peters, *Pitt and Popularity*, pp. 177–87.

[16] *Grenville Papers*, i. 379. The account and quotation that follow are taken from here, in subsequent Jenkinson letters, pp. 380–7.

main points at issue. The cabinet met on 19, 20 and 24 August to discuss the government's next move in an atmosphere of gloom and despondency. Pitt drew up the draft reply and quarrels developed over its hostile wording. These internal squabbles reflected the divide between the majority of the cabinet, still desirous of peace, and the obstinate aggressors, Pitt and Temple, now weary of parleying with the French. At the meeting on 24 August agreement was reached on a further peace initiative, allowing the French some limited access to the Newfoundland fisheries while remaining firm on the other points. Pitt and Temple dissented but refrained from a more dramatic show of temper. The whole exercise proved academic. In September Choiseul sent the proposals back with a clearly worded message that the war would continue. Worse still, Britain would probably be facing both France and Spain, in the future, for with Choiseul's rejections came the news that a Franco-Spanish treaty had been concluded.

This information shattered the uneasy calm that had existed in the cabinet over the summer. The Bourbon alliance vindicated Pitt's belligerent attitude towards the French, and he demanded that his colleagues endorse his plans for a pre-emptive strike against the Spaniards. They refused: as Walpole told Horace Mann, 'The Cabinet Council were for temporizing. This is not *his* style.'[17] The majority in cabinet supported Bute's argument that the government should first use the threat of war in the negotiation before opening hostilities. Isolated yet again, Pitt felt he had no option but to resign. At a cabinet meeting on 2 October his proposal for a Spanish strike was formally rejected. In the aftermath Pitt issued his threat of resignation which took place on 5 October, followed within a week by Temple and James Grenville. 'It is difficult to say which exulted most on this occasion', was Walpole's astute comment, 'France, Spain or Lord Bute, for Mr. Pitt was the common enemy of all three.'[18] With sound advice from Dodington, now Lord Melcombe, Bute certainly prepared the ground well for Pitt's departure. He realized there would be an outcry when

[17] *Letters*, v. 124.
[18] *Memoirs of George III*, i. 62.

Pitt's departure became known publicly, and came forward immediately with offers of alternative employment, and pensions and titles for his family. This softened the blow, and with the help of a press campaign to discredit Pitt's conduct, the government survived the winter of 1761–2 without much upset.[19] Bute's plans also included a permanent rôle for Grenville in a reconstructed administration. The problem for Bute in late September was to isolate George from the emotive scenes that would follow Pitt's resignation. He did not want Grenville to be carried off in a show of family unity with Pitt and his two brothers. On 25 September therefore Bute told Grenville in a carefully worded letter that 'though things look with a very doubtful aspect' he could see no reason for him to stay in town. If anything was resolved over the Spanish business, he added, Jenkinson would inform him in the country.[20] Grenville accepted the advice, probably glad of an opportunity to escape the constant political disputes within the family. On arriving at Wotton Grenville replied to Bute that he neither knew nor cared about the intrigues he left behind, 'which I have neither light nor discernment sufficient to see through'.[21]

Bute was not taken in by this self-deprecating tone, for he had a plan in hand to resolve Grenville's frustration. At two o'clock in the morning of 3 October an express arrived from Bute at Wotton, requesting Grenville to make his way to London at once. The letter retold the story of Pitt's declared intention to resign and indicated that Grenville would play a leading part in any new arrangement. 'The high opinions I have of you, the warm friendship I feel for you, and the entire confidence I place in you', wrote Bute, 'makes me see this dereliction with much more indifference than I otherwise should do.'[22] Grenville replied with a short note before departing for London, and it threw a shadow over the favourite's plans. Grenville agreed to come to town immediately, but he expressed little enthusiasm about the resignation: 'I have just now received your very kind letter

[19] Rea, *The English Press in Politics*, chapts. I and II.
[20] *Grenville Papers*, i. 388.
[21] Ibid., p. 389.
[22] Ibid., pp. 392–3.

giving me an account of the unpleasing and anxious event of Mr. Pitt's resignation, which I most sincerely lament, both as a public and as a private person'.[23] Bute's spirits must have dropped when he read this note, for the proposition he had in mind was to offer Grenville the seals to the secretaryship of the southern department just vacated by Pitt. The message from Wotton should have warned Bute that Grenville might refuse such an offer, and this is what he did on the afternoon of 3 October. The reasons Grenville himself gave for refusing the post concerned matters of family delicacy. There could well be some truth in this too. Ambitious though he was, Grenville did not want to be seen by his relatives, as both the instrument and executioner in Pitt's downfall. Further, in a fateful coincidence, Grenville had met Temple on his way to see Bute in the early hours of 3 October. What passed in Temple's post-chaise concerning Grenville's personal and political reliance on his elder brother may have convinced him not to accept the seals. A more likely explanation, however, is that all the old fears about separating himself from the political solidarity of the family and Pitt's protective oratory in the Commons had returned. Not only would Grenville have to acclimatize himself to a senior post of which he had no experience but also face Pitt in the House without any dependable friends or allies. Grenville detested Fox, potentially the ablest ministerial bulwark in the Commons, and refused to serve with him. His faint heart saw only isolation and persecution ahead.

What happened next lent weight to Grenville's case that he would be hopelessly isolated. After failing to persuade Grenville to accept the seals, the king accepted Grenville's suggestion of appointing his brother-in-law, Lord Egremont, to the southern department. But Bute did not give up hope of pushing Grenville into a similarly important role in the reconstructed administration. From the account in his wife's diary it is clear Grenville believed that he would be allowed to slip quietly into the speaker's chair: this request, however, was politely refused. If the king and Bute were to create the administration of their choice, Grenville's ability as a man of

[23] Bute MSS 3/25/1.

business in the Commons would be at a premium and must be acquired. The first plan to achieving this end was vetoed by Newcastle in late September. It consisted of making Grenville chancellor of the exchequer and leader of the House of Commons. Newcastle did not want to change a treasury which he controlled, nor exchange the present docile chancellor, Barrington, with someone far more knowledgeable and able in financial affairs, who already had two close friends, James Oswald (MP for Dysart) and Gilbert Elliot (MP for Selkirkshire) on the treasury board. This Newcastle saw as an open invitation for undermining his position from within and refused to countenance the changes. Thwarted on this front, Bute and the king came up with what was to be the final solution, promoting Grenville to leader of the House, with a seat in the effective cabinet while retaining his position as treasurer of the navy. This was power without the appropriate office to make it secure. The plan was put to Grenville on 12 October, and expressed more fully in a letter from Bute the day after.[24] The letter had to be returned immediately, however, because of several disparaging remarks about Newcastle. So before making a final decision on the offer, Grenville dictated to his wife recollections of its principal contents. The core of the argument, as Grenville saw it, represented a demand for loyalty in perilous times. To save the king from Newcastle and his friends, or even Fox should matters deteriorate further, Grenville had to come forward and carry the king's banner in the Commons. Sir Lewis Namier has analysed Grenville's recollections at length, comparing them with Bute's original letter.[25] He concluded that in his way Grenville laid the foundation for a break with Bute by avoiding reference to the favourite in crucial passages of these recollections. Yet, in this critical hour it is difficult to believe that Grenville possessed such purpose of thought and deed. Wracked with indecision and self-doubt, these notes dictated to his wife were a basic exercise in confidence building. To convince himself that he was not 'being led into a situation amongst people hostile to me', Grenville needed

[24] *Grenville Papers*, i. 395–6.
[25] *House of Commons*, ii. 538–9.

the assurance the king 'would support me to the utmost' and that 'my honour was his honour, my disgrace his disgrace'. Bute's letter obviously gave this assurance and after weighing up the arguments Grenville agreed to lead the Commons in the new parliament.

He held no illusions about accepting the post. The personal cost was high. All his relatives, except Thomas Pitt, turned against him. Temple went so far as to cut George's sons from his will, and Pitt insisted that Grenville no longer oversee the finances of his wife. The pathetic letters between George and his sister Hester, now Lady Chatham, to arrange the details of this transaction express Grenville's grief that political differences had been allowed to divide the family in such a way.[26] The parliamentary situation did not look any rosier. Though he looked forward to the opening of parliament in November with the full weight of royal patronage behind him, his position in the Commons would be an unenviable one. Grenville had no following of his own or the means of acquiring one. Bute did not delegate any powers of appointment or dismissal, leaving Grenville with no real means of cracking the whip over the government's supporters. Worse still, he would have to face Pitt's attacks in the Commons reliant on the friends of Newcastle and Fox over whom he held no sway. Indeed, when Bute secured Fox's support for Grenville, it merely emphasized the uncertainties facing the new leader of the House.

> What figure shall I make? [Grenville asked Newcastle] Mr. Fox has superior parliamentary talent to me; Mr. Fox has a great number of friends in the House of Commons, attached strongly to him; Mr. Fox has *great connections*, I have none; I have no friends; I am now unhappily separated from my own family.[27]

Bute did his best to allay Grenville's fears, but it did not prove an easy task. By the beginning of November Bute had been driven to the point of distraction with requests for declaration of support and confidence. In three weeks Grenville's moods changed from confidence to utter self-doubt, when at one point he suggested that Lord Barrington conduct the eve of session meeting for government supporters

[26] See, for example, Grenville to Lady Hester Chatham, 4 Dec. 1761, B.L. Add. MSS 57807, fo. 62.

[27] Cited in *House of Commons*, ii. 539.

at the Cockpit. Bute did finally bring Grenville around with a firm and patient campaign of reassurance. By early November Bute was telling all senior members of the government that he expected their full support for Grenville and that attention be paid to his recommendations.[28] As a declaration of intent on this point, the king granted Grenville's request that his brother-in-law, Lord Thomond, be made cofferer and Sir George Savile comptroller of the household.

Bute's labours proved well worth while. Once Grenville assumed his new rôle the fears and doubts began to evaporate. In policy matters he found himself at one with Bute and Egremont, and it was Newcastle who began to feel the outsider in the cabinet. This applied in particular to the question of Spain's continued intransigence. Though Pitt's idea of a pre-emptive strike had been dismissed, the peaceful overtures that followed failed to produce any progress towards a peaceful solution of the claims against Britain. In fact by the end of November the situation had approached the point where a decision over whether to open hostilities or not would have to be made; with Bute, Grenville, and Egremont leading those in the cabinet favouring a show of force. The Cockpit meeting on 1 November provided a further boost to Grenville's confidence. In this eve-of-session gathering for the government's supporters Grenville showed that he could deal competently with objections from the large body present over wording in the Address.[29] This was perfect preparation for the opening day of parliament on 13 November. Many MPs expected a high-spirited attack on the ministry from Pitt, but he remained moderate in his statements on the Address. Two matters were raised during debate: first, the omission of any reference to the expiring Militia Act in the Address, and second, on Pitt's initiative, a request to 'lay open the proceedings of Spain' to allow the House a word in what action the government should take next to settle the dispute with the Bourbon powers. No one doubted that Pitt's motives in this were dictated by a desire to embarrass his former colleagues, and justify his strong line on the question in September. Grenville declined

[28] *Bedford Correspondence*, iii. 67.

[29] Walpole, *Memoirs of George III*, i. 68. The account of the debate on the Address on 13 Nov. and quotation that follow are also taken from here pp. 63–7.

the request by a deft sidestep: he took note of all the issues raised in the debate, he told MPs, but their omission arose from a point of procedure. Each issue could come before the House at a later date. Grenville had just done enough not to upset Pitt and raise unnecessary difficulties. As Lord George Sackville observed, he made 'no declarations of what might hereafter be the measures so as to give anybody a handle for fixing him down to any particular system'.[30]

The luxury of such equivocation had to be abandoned in the following month when more thorny issues came to the fore. Securing an elusive peace and renewing legislation for the militia caused the government some uncomfortable moments in the weeks before the Christmas recess. The government's dilemma over the search for a peace first became apparent in debate on 9 December. On this day discussion took place in committee on the payment of the Prussian subsidy for the coming year. The majority view in cabinet favoured an end to these huge payments which Charles Townshend, secretary at war, estimated at £5m for the coming year. Yet, as long as the war continued, how could this be done without revoking Britain's alliance with Frederick or rescinding firm promises to maintain material support for this front? No one to that point had come up with the answers. The Court supported war in Germany, wrote Walpole on 12 December, 'for they don't know how to desert it'.[31] Grenville was no exception. In a defensive speech about the subsidy and the war in general he made public his long-held reservations about Pitt's expensive continental policies. He

> supported the question solely on the foot of treaties, but took care to assert that he had neither advised nor approved them; he had not been able to stop a torrent . . . An immense load of debt had been laid upon us: he would not call on any light of Government, who had brought us into these distresses, to help us out of them . . . our honour was pledged, and he would not be for an igominious peace; nor on the other hand would he intoxicate the people with unattainable objects.[32]

The speech itself hardly represented a masterpiece of oratory. There was little credit for Grenville in saying he opposed the

[30] H.M.C. Stopford-Sackville, i. 86–7.
[31] *Letters*, v. 152.
[32] Walpole, *Memoirs of George III*, i. 81.

subsidy policy in the previous administration while remaining in place and avoiding publication of his reservations. As a declaration of war against Pitt, however, it was the perfect beginning and a transformation from the self-doubting figure of two months earlier. On this performance, Grenville would be leader of the House in word and deed, and provide a rallying post for those MPs genuinely desirous of peace but too timid to disassociate themselves from Pitt's '*divine plan*'. Bute was unstinting in his praise. 'Millions of Congratulations upon your very great, very able, and manly performance', he wrote on 10 December, 'this will do, my dear friend, and shows you to the world in the light I want, and as you deserve.'[33] Though these comments had the air of a headmaster praising his head boy at school, Grenville no longer doubted that he had the full weight of royal support behind him.

In this debate Pitt had reacted to Grenville's unexpected denunciation of his policies with disdain, refusing even to direct his answers justifying the German war at the new leader of the House. This response showed that Pitt realized Grenville had taken the initiative, and that he and not Pitt had captured the sympathies of members present. This point came home at the conclusion of the debate when Pitt had been subject to a violent character assassination by Isaac Barré, the newly elected MP for Chipping Wycombe. One observer summed up the reaction of the House to such a rare occurrence: 'Col. Barry's attack, tho' rough and indecent on Mr. Pitt, will it's thought, have no bad effect; and, as he never spared others, is the less regretted.'[34] The demolition of Pitt's position continued two days later on 11 December when he supported a motion requesting papers on the Spanish claims against British fisheries be laid before the House. These papers had been mentioned in the debate on the Address, and George Cooke (MP for Middlesex) explained his reasons for the motion, 'by saying he wished to know what was the state of our affairs with regard to Spain – he hoped, peaceable; but desired to see the papers relating to their claims, and to know if they treated us with contempt

[33] *Grenville Papers*, i. 418.
[34] *Caldwell Papers*, II. (i), 137.

and disdain'.[35] No one took this seriously. Calling for papers was a traditional smoke-screen for condemning the actions of a previous administration, led by Grenville, ministers resolutely denied the request. Grenville put the government's case in a nutshell:

> that though he had heard of this motion he could not believe it would be made. It was our right to have papers, when essentially necessary; but the power of negotiation belonged to the Crown; and negotiations ought not to be made public, when real mischief might be the consequence.[36]

This sensible view found support from all sides of the House, 'even his [Pitt's] friends the Tories, who had been falling back to him, abandoned him on this motion', wrote Walpole after the debate.[37] Yet the most remarkable aspect of the debate was again the intensity of the personal attacks on Pitt not only by Barré but also Gilbert Elliot. Their willingness to take on the role of antagonist marked the real turning-point in the struggle between those for continuing hostilities and the growing number of MPs in favour of peace. Walpole's account sums up the mood of the House perfectly in the light of these attacks: 'Pitt made no manner of reply', he wrote, 'It seemed to some a want of spirit, but it was evident from the indignation of the House, that such savage war was detested.' It was not often that Pitt declined a slanging match, but it proved symptomatic of the day's proceedings. Pitt had come to the House confident of forcing a close division, in the end the motion was rejected with 'scarcely six voices being given for the question'.[38]

Grenville's reaction to this result could only have been utter delight and satisfaction. Drawing Pitt's venom provided an unexpected bonus to the prime task of establishing himself as an effective leader in the House. The extent of his success can be measured by the small number of demoralized opposition MPs who took part in the Militia Bill debates the following week. Renewing the militia laws, due to expire within that parliamentary session, had promised to be a

[35] Walpole, *Memoirs of George III*, i. 88.
[36] Ibid., p. 89.
[37] *Letters*, v. 152.
[38] Walpole, *Memoirs of George III*, i. 95–6. See also *Border Elliots and the Family of Minto*, pp. 371–2.

tricky problem for the government since they were brought up on the day of the Address. The point at issue was merely duration.[39] Grenville himself favoured a straight vote for perpetuity. It was not wise for the government, in his view, to be seen bickering over limitations when a threat of invasion from France existed. Major revision could be undertaken in the cold light of day when a peace had been secured. Opinions within the ministry were divided, however, with Newcastle and Bute representing the majority who favoured a set limit of years on any renewal. After considerable lobbying this view prevailed and would be presented to the House as ministerial policy. At a large meeting of opposition MPs at the St Albans tavern on 25 November 1761 a similar difference of opinion arose. But in this case, urged on by Pitt and his lieutenants, the meeting resolved to move for perpetuity and take on the government to secure this end.[40] This set-piece battle should have taken place on 15 December when Lord Strange presented the new militia bill to the Commons. Instead the government robbed Pitt of the opportunity of capitalizing on the measure by deferring any further proceedings on the bill till after Christmas. The successes achieved earlier in the session were now complete. In the long committee stages revising the content of many clauses in the new bill, Pitt found it quite impossible to raise a storm against the government's handling of the measure. By 19 March 1762 the ministry's grip on the legislation was so firm that the clause limiting the new bill to seven years' duration passed the House without a division.[41]

Due to the worsening relations with Spain, Grenville savoured victory for no more than two weeks. The government failed to match its successes at Westminster in the diplomatic field. In late December a flurry of activity and exchanges between Britain and Spain over the latter's claims in North America ended in deadlock; and on 2 January the cabinet issued a formal declaration of war against the Spanish. For two principal reasons no one in the ministry relished a new campaign. First, it endangered loans secured

[39] For a fuller discussion of the issue see Western, *The English Militia in the Eighteenth Century*, pp. 184–93.

[40] Walpole, *Letters*, v. 151.

[41] *Commons Journals*, XXIX. 249–50.

by the Duke of Newcastle in the City for the coming year. Though no binding undertakings were given, money had been raised on the understanding that the government would be cutting its expenditure on the war, especially in Europe. Second, it gave ample opportunity for Pitt to gain some sort of revenge on his former colleagues. He and his supporters could not justify their own conduct and policies during the summer of 1761, and the government knew there were sure to be demands for the restoration of Pitt to direct the strategy he initiated six months earlier. Ministers felt confident that they could handle the latter criticism but the question of expenditure proved more difficult. John Crauford (MP for Berwick) described their dilemma and projected solution:

> disunion or rather discontent among the Ministers, is supposed, from appearances; and some think our friend [Bute] should stand forth and be more determined; Spanish war not dreaded, tho' it sinks credit; a strong wish and hope the Ministry will lessen our expense and exertions in Germany, and that events will arise to show that we neither can nor ought to support where it's impossible to do good.[42]

The discontents that Crauford mentioned in his letter referred specifically to a dispute between Newcastle and the rest of the cabinet over the hope to lessen expense and exertions in Germany. Newcastle had championed the subsidy treaty and the Prussian alliance throughout the war, and to talk now of reducing payments to Frederick in order to finance the action against Spain undermined the whole thrust of his diplomatic strategy. The brunt of his anger fell on Grenville who was the most active supporter in cabinet of the Spanish campaign favoured by Bute and the king. In the first three months of the year relations between the two hit rock-bottom. There were many facets to Newcastle's anger and dissatisfaction with Grenville; some were personal, others the product of Newcastle's dismay and frustration at being rejected by the Court. That Newcastle did have grounds for complaint with Grenville, however, there can be no doubt. Following the successes against Pitt in December Grenville's confidence and standing within the ministry had grown considerably. Bute and Egremont turned to Grenville for advice, deliberately cutting Newcastle out of the

[42] *Caldwell Papers*, II. (i), 139.

preparations for the Spanish war and the reduction in subsidy to Frederick. This was impertinent behaviour towards someone who had served the public for almost forty years, and was made worse by Grenville's insistence on interfering in treasury business. As leader of the House, Grenville was entitled to have access to departmental papers, but he exploited this right for his own ends. With the co-operation of his two friends at the treasury board, James Oswald and Gilbert Elliot, Grenville began a systematic campaign to discredit Newcastle's subsidy policy, prompting the duke to complain to Devonshire in April that he dared not write to his fellow secretaries any more for fear that it would be 'scanned by Mr. Grenville'.[43]

The dispute came to a head in April over the government's intended application for £1m credit to support the war in the ensuing year. Newcastle believed that the money should be applied to the continental campaign, and the Spanish war allowed to splutter to a close. To fight on both fronts would, in his opinion, necessitate the raising of a further million in the coming year. Grenville, Bute, and Egremont took the opposite view, promoting the idea that the £1m could serve both fronts if the government reduced the Prussian subsidy. Grenville certainly prepared the ground well for disputing his case as the argument developed. As Newcastle told Hardwicke 'Mr. Grenville engaged part of my own Treasury against me, and quoted to me their opinions; . . . they not only quoted them against me but sent regular questions.'[44] Gilbert Elliot was one that Grenville engaged against Newcastle; and amongst his papers there is a draft copy of written and statistical evidence supporting the case for the Spanish war at the expense of the continental campaign.[45] It was written in late March 1762 and echoes to the letter views expressed by Grenville to Jenkinson in early April. In this note Grenville reflects on the government's decision in March to reopen secret negotiations with France, and reaches the conclusion that Newcastle's policies would ruin any chance of a settlement:

[43] Yorke, *Hardwicke*, iii. 347–8.
[44] Ibid., p. 356.
[45] Minto MSS 11040, fos. 40–43.

> nothing can contribute so effectually to peace as to convince France by a great reduction of our expenses that it is in our power though not in our wish to continue the war, as nothing but the contrary persuasion can induce France and Spain in their present circumstances to go under such heavy and repeated blows.[46]

Outmanoeuvred on every front, Newcastle was compelled to put these differing views on strategy to the vote. This took place on 27 and 30 April when the cabinet met to discuss the vote of credit, and Newcastle suffered ignominious defeat. Only Hardwicke and Devonshire supported him. On 7 May Newcastle told the king of his resignation and on 25 May it was formally announced, bringing to an end a career of thirty-eight years in public service.

Grenville's victory was completed when he spoke 'finely' for the vote of credit in the Commons on 12 May. Even with Newcastle on the way out, the opposition led by Pitt could make no impression in the government's support. To show their disdain for the opposition attack, MPs voted the money without a division.[47] First Pitt and now Newcastle had fallen victim to Grenville's growing prestige. Bute too soon found himself at odds with Grenville s ambition for high office when remodelling the ministry on Newcastle's resignation. The initial wish of Bute and the king was for Grenville to become chancellor of the exchequer at a treasury with Bute as first lord. This suggestion fell down due to Grenville's refusal to accept the office. His reasons were blunt and to the point. As the king told Bute tetchily on 6 May, 'I am inclined to think that Grenville is weak enough to think he may succeed the Duke of Newcastle and not just be an assistant in a Board.'[48] Bute in his turn resisted Grenville's arrogant proposals with a sharp appeal to his colleague to honour his previous declarations about the exchequer, which he 'had ever looked upon as fixed'.[49] It proved to no avail. Grenville had his eye on Newcastle's vacant post at the northern department and pressed the promotion with Bute and the king *ad nauseam*. In the middle of May Bute told him that the chances of this being carried were remote, for

[46] *Jenkinson Papers*, pp. 39–40.
[47] Walpole, *Memoirs of George III*, i. 127–8.
[48] Sedgwick, *Letters from George III to Bute*, pp. 100 and 104–5.
[49] *Grenville Papers*, i. 446.

'two brothers, secretaries, having the same office will not be relished'.[50] He saw a compromise formula in which Egremont went to Ireland as lord lieutenant and Grenville took the seals alone; but the latter reacted violently to the suggestion, 'refused to take any part in this proposition, and said it was impossible for him to take the seals, to have a personal affront given at the same moment to Lord Egremont'.[51] Unable to conjure up any further permutations Bute and the king gave way: on 28 May Grenville took the seals as secretary of state for the northern department. The family biographer, L. M. Wiggin, has enumerated the qualities Grenville used in reaching high office thus, 'hard work, ability and at the last stubbornness'.[52] This paints the picture of an honest plodder reaping his reward for good behaviour, and though this limited view is not without foundation, the portrait as a whole is incomplete. Since his promotion to leader of the House, Grenville has displayed an incisive streak in his character hitherto unseen. He had been shrewd and clever in the campaign to oust Newcastle, and cool-headed when the duke began to fight back in the cabinet. There is no doubt that he had also made his weight felt in cabinet through the force of his personality and intellect. At the time of his retirement Newcastle did not over exaggerate when he said that Grenville had both Egremont and Bute under his influence, for he read their correspondence and wrote many of their reports and speeches. The qualities he exhibited in these months, in fact, stood Grenville in good stead when he eventually became first lord the following year.

Unfortunately the experience of high office did not fulfil the expectations raised in the struggle to attain the seals. Grenville spent five unhappy months as northern secretary. In an office where he should have been able to build on the prestige and influence gained since the accession of George III, nothing seemed to go in his favour. In most studies of this period the root cause of this failure is put down to a simple mistake by Bute. As Fitzmaurice, the biographer of Shelburne, wrote in 1912, Grenville was promoted to an

[50] Ibid., p. 447.
[51] Ibid., p. 450.
[52] *Faction of Cousins*, p. 264.

office 'for which he was not fit'.[53] Though the claim has often been repeated, no further explanation is usually offered, with the implication being that Grenville could not cope with the demands of conducting foreign policy.[54] This view hardly bears examination. Grenville may not have dealt directly with ambassadors before but he had had considerable experience of diplomatic policy and correspondence over the previous year. Bute certainly relied on his judgement in these matters. Shortly before Newcastle's resignation, for example, Grenville had written a long paper for him on the request from Portugal for assistance against Spanish aggression.[55] The only interpretation that makes sense of the view that Grenville was not fit for this office relates to the fact that during the summer his opinions on the quest for peace differed from those of Bute and the king. This made the chances of agreement in cabinet less likely and friction more probable.

On the surface there appeared to be little difference between the views of Bute and Grenville on Britain's aims in ending the war. They shared an overriding desire to reduce the burden of debt by an early peace with France and a reduction of expenditure on the German campaign. This was the policy initiated at Leicester House, and the fundamental dividing line between their approach and that of Pitt. On the German issue Bute and Grenville remained in accord throughout the summer. Despite intense recriminations from Prussia, they bore the responsibility for terminating the treaty guaranteeing Frederick's subsidy, which was to have such far-reaching effects on Britain's European alliances over the next twenty years. The negotiations with France proved a different matter. By May 1762 Bute's attitude to the old enemy had mellowed considerably. His desire for peace was stronger than ever, and he favoured fresh territorial concessions to achieve this end. The reasons for this change of heart are not hard to find. There is no doubt that Bute and the king had long tired of the deadlock over terms and wished to deliver the peace pioneered on Pitt's resignation

[53] *Shelburne*, i. 108.

[54] See, for instance, Sir Lewis Namier's short biography in *House of Commons*, ii. 539.

[55] *Grenville Papers*, i. 440–43.

almost a year earlier. It seems certain too that Bute's resolve suffered erosion under the growing press campaign against his promotion to first lord. He would have been inhuman not to feel a side wind from some of the scurrilous stories printed about him. There was a need on Bute's part to complete his self-appointed mission as peacemaker in order to justify the ousting of Pitt and Newcastle, and, more important, give himself a chance to escape from the limelight for a while.[56] Whatever moved Bute had no effect on Grenville, who could not agree to the concessionary policy. The first note of acrimony that occurred between the two concerned Guadeloupe. At a cabinet meeting on 23 April ministers discussed the restoration of this island with Martinique in return for territory on the American mainland. Grenville opposed the restoration proposals outright as inadequate compensation had been offered. The meeting broke up without reaching agreement, but at a further gathering the following week, when Grenville lay ill in bed, Bute secured cabinet support for the restoration of Guadeloupe. Grenville took this as a personal affront, and in the first week of May he explained to Bute the basic principle that governed his attitude to the French negotiation. To give any succour to Choiseul, France's chief negotiator, as had been done in 1761, he wrote, 'would carry us through the same track of ruin once more'.[57] In his diary Grenville went further than this, accusing Bute of losing his nerve. As he saw it, Bute only proposed fresh concessions because he 'feared the negotiation might break off'.[58]

In this atmosphere of distrust and disillusionment relations between Grenville and Bute cooled rapidly. To Grenville, Bute's weakness, exemplified in his conduct of the negotiations with France, endangered the country's future security and negated many of the sacrifices made by British forces during the war. Grenville's vision of Britain's destiny was at this stage blurred, but at its core there existed a conviction that all resources should be concentrated on expanding the seaborne empire. This expansion could be secured at a reasonable

[56] Brewer, 'The Misfortunes of Lord Bute: A Case Study in Eighteenth Century Political Argument and Public Opinion', *H.J.* 16 (1973), 3–43.

[57] Bute MSS 3/28/2.

[58] *Grenville Papers*, i. 450.

cost by the British navy, and wasteful subsidies to tie up French forces on the Continent avoided. In America and the West Indies Grenville aimed to deny the French any foothold which could provide an excuse for future hostilities or be a serious threat to British trade. Here he parted company with Bute whose restoration policy would, Grenville believed, undermine the strategic advantage gained over France during the war. This view of the future represented a mixture of the Tory country-war policy and old-fashioned Whig pragmatism that blended perfectly with the war-weary mood of the country. Concern over the vast increase in the national debt and the lack of visible means to reduce it, became a national obsession in the early 1760s. Grenville's message of reducing expenditure and debt without damaging Britain's supremacy in the New World offered an instant panacea for these anxieties.

Needless to say Bute and the king did not see the peace settlement in just this way. In their eyes, Grenville's enhanced self-esteem and popularity in the cabinet no longer represented a useful weapon with which to tackle Newcastle and Pitt, but an encumbrance that caused perennial difficulties over the terms of peace. Bute's initial reaction to this obduracy was to drive a wedge between Grenville and Egremont. In late June he persuaded Egremont to sign a secret dispatch to France offering the unconditional cession of St Lucia in return for Choiseuil's approval of the other British proposals. No one else in cabinet knew of this proposition when the French delivered their formal acceptance to the king on 2 July, and Bute believed that he had outflanked his internal opponents. At a cabinet meeting on 26 July Bute presented the French acceptance and the St Lucia clause as a *fait accompli*, requesting that his colleagues merely rubber-stamp the finished article. He received a rude awakening, for, led by Grenville, the majority of ministers at the meeting vehemently opposed the settlement. They expressed anger not only at the terms offered to France, which one of Grenville's officials had described as 'very lame and insufficient' but also Bute's willingness to settle without Spanish approval or involvement.[59] The

[59] *Grenville Papers*, i. 464.

consensus of opinion was that the war could go on indefinitely unless both Bourbon powers put their signatures to the peace, and giving away hard-won war prizes in the West Indies was not the way to bring this about.

This internal revolt shocked Bute and the king, but they refused to be bullied into making the tough stand advocated by Grenville. As a declaration of faith in the proposals agreed between Bute and Choiseul, they decided to press on and send the Duke of Bedford to Paris to conclude the final articles of peace. It proved a provocative choice. Bedford, a haughty and proud man, made no secret of his enthusiasm for the terms already agreed or his dislike for the position of Grenville and his colleagues. Unfortunately, these feelings were reciprocated by Egremont and Grenville, who had no faith in Bedford's ability as a negotiator. Viry, the Sardinian envoy in London, wrote of their attitude 'Il n'est pas mal avec Milord Egremont . . . Pour M. George Grenville, il le méprise et ne le croit point propre pour le Poste qu'il remplit.'[60] In his determination to see the peace initiative to a successful conclusion, Bute also resolved to bring into the administration someone powerful enough to counter Grenville's influence in the cabinet. Thus at the end of July Bute approached Newcastle through Hardwicke, intimating that if he would do the king's dirty work in cabinet he was free to name the office of his choice.[61] The overture was dismissed out of hand, for by this time Newcastle held the same low opinion of Bute's double-dealing as Grenville. Bute accepted this rebuff philosophically, and made no further attempts to change personnel in August and September. For solace he knew that if Bedford concluded an acceptable and beneficial peace treaty, Grenville's guerrilla campaign in the cabinet could be overcome. Grenville had no such comforting thoughts as he retired to Wotton for the remaining summer months. Only doubt lay ahead. He could engender no enthusiasm for Bedford's mission which began on 6 September, and realized his place was in jeopardy. 'This difference of opinion between Lord Bute and him', wrote his wife, 'gave grounds to his enemies to work with greater

[60] Cited in Rashed, *The Peace of Paris*, p. 167.
[61] O'Gorman, *The Rise of Party*, pp. 40–41.

success than they had hitherto done.'[62] In fact there was an irony in Grenville's isolation, in that his stubborn resistance to the policies of Bute and the king so resembled Pitt's position a year earlier. Bute even suspected that 'Grenville was not so much at variance with his family as he wished to be thought'.[63] This imputation proved groundless, and yet these seeds of suspicion took root as the summer wore on and the misunderstandings multiplied.

The bickering and in-fighting recommenced in the last week of September when Bedford's first dispatches arrived from Paris. On 24 September Egremont sensed that trouble might be brewing and pleaded with Grenville to 'come to town soon; I fear you will be much wanted'.[64] His fears seemed well founded when the overdue packet of papers arrived two days later. Bedford had, in Egremont's view, predictably overstepped the mark in the negotiation, ignoring his instructions not to sign or agree any article without prior consultation with London. He sent the dispatch to Grenville, who was still in the country, commenting that, 'you will see that headstrong silly wretch has already given up two or three points in his conversation with Choiseul, and that his design was to have signed without any communication here.' Bute was not impressed with Egremont's criticisms of the dispatch. Two further packages arrived on 27 and 29 September, and in these the final articles of peace began to take shape, prompting Bute to express confidence that the negotiation would succeed. On 29 September, however, the whole process became uncertain with the news of Havannah's capture by a British task-force under the Keppels. This previously impregnable fortress represented the greatest prize of the war, and no peace would now be possible with Spain until compensation for its restoration had been decided. Indeed, within one hour of the news breaking, Grenville and Egremont declared they would not sign peace which did not allow of compensation being agreed for the capitulation of Havannah. This put Bute on the spot, for Grenville had undoubtedly taken a stand that would prove popular in

[62] *Grenville Papers*, i. 450.

[63] Walpole, *Memoirs of George III*, i. 154.

[64] *Grenville Papers*, i. 474. The account and quotation that follow are also taken from here, pp. 475–81.

parliament and the country at large. At a cabinet meeting on 4 October therefore Bute tried to persuade his colleague not to endanger the whole negotiation by insisting on an immediate solution to the Havannah problem. Though Grenville was absent, it made little difference to the determination of the cabinet to support his tough line on the question of adequate compensation. Feelings ran so deep that Bute was obliged to change his mind on the issue, and the following week Bedford received instructions to seek compensation for the return of Havannah.

It is often assumed that the dispute over Havannah, coming so soon after the St Lucia incident, was the reason behind Bute's decision to remove Grenville from his post. In fact it was no more than a catalyst. In late September and early October a more fundamental difference of opinion had arisen over leadership in the Commons. When Bute turned his thoughts to the task of piloting the peace through parliament he foresaw problems on two points. First, he intended to lay the preliminaries before both Houses in their finalized state. Rather than discuss the various articles, MPs would merely be asked to approve or reject the settlement. There was nothing untoward in this: Gilbert Elliot prepared a long paper on the precedents to support such a move, concluding with the comment that 'the king's prerogative undoubtedly empowers him to conclude peace without laying terms before parliament'.[65] If trouble arose over the compensation for Havannah, however, MPs could claim that Bute was riding roughshod over their privileges on a matter of such importance to the nation's future. On delicate constitutional issues like this, the outcome was never certain, and difficulties gaining approval for the settlement as a whole might be experienced. Second, Bute harboured serious doubts about Grenville's ability to guarantee a Commons' majority for the peace. He knew that Grenville lacked real commitment to certain articles of the peace settlement, and it showed in his attitude to the coming session. Grenville had still not returned to London by the second week of October nor begun lobbying MPs for the opening of parliament due in November. Thus Bute was

[65] Minto MSS 11036, fo. 26.

naturally prompted to look for someone upon whom he could depend to carry out his wishes to the letter. On 6 October he sent Shelburne to see Henry Fox with a royal summons, and the old campaigner agreed to take on the role of executioner. Grenville did not feel that he had to justify his ambivalent behaviour towards Bute. His reticence to come to town and organize the government's troops made perfect sense when he was trying to force Bute's hand on both issues in dispute. On the first point, unlike Bute and the king, Grenville believed that the government should show respect to the House by laying the preliminaries before it before they were signed. This policy had a double edge. Grenville, on the one hand, could argue that he was simply observing normal parliamentary procedures while on the other, he knew that laying these clauses open to discussion would exculpate him from any blame in the public eye for the unpopular articles granting concessions to the French. On the second charge Grenville's defence was more clearly defined. As he told Bute in September, he could no longer carry out his duties as leader of the House without the substance as well as the appearance of power. No serious canvass could take place unless Bute authorized him 'to talk to the members of that House upon their several claims and pretensions'.[66]

In the wake of this challenge to his leadership, Bute determined that Grenville, with his scruples about the peace, must be removed from his senior post. The demand that the preliminaries be laid before parliament for discussion was sufficient in itself to warrant a change: to talk of relinquishing his power in the Commons went beyond all reason and was something Bute would never consent to.[67] On 10 October Bute interviewed Grenville to inform him of his decision to bring in Fox, describing it as an 'expedient, to obviate the difficulties likely to arise in Parliament'.[68] Grenville's talents were not to be abandoned completely, Bute said, 'though the king found himself obliged to do this, he hoped it would be but for a time: that his Majesty hoped Mr Grenville

[66] *Grenville Papers*, i. 483.
[67] *Bedford Correspondence*, iii. 135 and *Grenville Papers*, i. 483.
[68] This and the quotation that follows are taken from Grenville's own account of his demotion, *Grenville Papers*, i. 451–2.

would still continue in his service; that he intended to make Lord Halifax secretary of State, and Mr Grenville First Commissioner of the Admiralty.' After a night to recover his composure at hearing this news Grenville then had a similarly unpleasant interview with the king. George III regretted the changes but told Grenville that '"we must call in bad men to govern bad men"'. There was nothing in these explanations that expressed the true feelings of the protagonists towards Grenville's demotion. Grenville had gone into these interviews, as Fox told Bedford 'half unable, half unwilling' to continue to support the conduct of the peace negotiations, and thus desperate to know whose advice would be taken. The king took the side of Bute without hesitation. Both were tired of Grenville's monotonous objections to their policy, and concluded that it was impossible for him to take responsibility for measures in which he had no faith. In the short term their plan proved an undoubted success. The cabinet swung behind the preliminaries negotiated by Bute, and Fox carried the peace to the Commons without a dissenting government voice. Further, in simply moving Grenville from one office to another, and holding out the vague hope of promotion in the future, they retained his loyalty. By this point it was somewhat strained, yet Grenville did not seek out Pitt to concert a campaign against the preliminaries after his demotion. He took the admiralty, ate humble pie and nursed his grievances in the hope of settling the score at some future date.

In view of Bute's consummate mastery in his handling of these changes, what had prompted Grenville to make this bold stand that ended in such disaster? He did not take inspiration from his family and former associates, as Bute at first thought. His decision to accept the admiralty rather than resign on principle bears this out. Grenville needed a ministerial salary to support his family after Temple cut off any hope of private income in October 1761. Grenville even doubted that Temple would allow his return for Buckingham when he sought re-election on taking the new post. However, Temple did not think it worth interfering as the ministry could easily find Grenville a new seat. The humiliation that Grenville felt at his demotion is well summed up in Newcastle's letter to Devonshire on

14 October, when he asked how the ministry could 'venture to use Mr. Grenville so ill as to take from him an employment of the first rank and confidence of £8 or 9,000 to put him at the Admiralty with barely £2,000'.[69] At root Grenville was moved to take on Bute because he thought the favourite would be forced to give in to his demands. He had fought Bute over the terms of peace and gained some successes, why could the same rule not apply to the management of the Commons? Grenville's refusal to take responsibility for the preliminaries without the powers of a first minister came very close to the opening of the new session. It had the appearance of a well-calculated gamble, for to whom could Bute turn at that late hour if Grenville refused to go on? Pitt was out of the question, and, as Grenville told the king in his interview on 11 October, the thought that Fox might be brought back never entered his head,

> Mr. Grenville entered his protest very strongly against the step the king was going to take, stated the improbability of facilitating his affairs by calling in so unpopular a man as Mr. Fox, and foretold the ill success which must attend so desperate a measure: that as he himself had been called upon by His Majesty to Mr. Fox's power, that he had obeyed him by sacrificing to his commands a situation of ease, profit and honour.[70]

In some respects Grenville can be forgiven for this error of judgement. The king disliked Fox, and Bute's decision to recall him 'was the turning-point in the king's friendship with Bute'.[71] Fox was also unpopular at Westminster and in the country at large. 'Unpopularity heaped on unpopularity does not silence clamour', commented Walpole on hearing of Fox's promotion, 'Even the silly Tories will not like to fight under Mr. Fox's banner.'[72] Grenville can not be excused, however, for underestimating Bute's powers of persuasion over the king, of which he had ample experience. He went to the admiralty a sadder and wiser man, having paid a high price for questioning Bute's authority in this sensitive area.

The next six months Grenville spent in purgatory. Though he was still a member of the cabinet and attended meetings

[69] Cited in Wiggin, *Faction of Cousins*, p. 275.
[70] *Grenville Papers*, i. 451–2.
[71] Brooke, *King George III*, p. 95.
[72] *Letters*, v. 265.

regularly, his wings had been clipped. As Fox told Bedford on 26 October, the conflict backstairs had ended; 'Mr. George Grenville, though of the Cabinet, will not longer have it in his power to guide and interline Lord Egremont's drafts.'[73] Grenville did have the work at the admiralty to occupy him, but even here he could not escape Fox's long reach. In early November, for example, preparations were underway for Sir Charles Hardy to lead a squadron to intercept a French force under de Blenac attempting to return to Brest. Fox thought that to proceed with the mission as peace approached might be deemed provocative by the French, and he advised Grenville to consult Bute and the king. This he did and they supported Fox's argument wholeheartedly, inflicting yet another humiliation on their demoted secretary. Grenville accepted the rebuff in good heart because he had the last word. In a defiant statement that paid tribute to his application and understanding of official business, Grenville told Fox on 11 November that he was still in the right to continue with the mission,

> I cannot help observing to you that, by the Preliminary Articles, hostilities are to cease after the ratification of the Preliminaries, which the Council were yesterday unanimously of opinion must be construed after the exchange of the ratifications; and consequently, until it is signified to us from the Secretary of State, by the king's command, the Board of Admiralty cannot take any notice of them, or put a stop to hostilities at sea.[74]

This was certainly not the talk of someone who expected to remain in the political wilderness indefinitely. Nevertheless Grenville had to suffer more agony as the peace went through parliament before his chance for revenge arose. Fox took full command of preparing the ground for laying the preliminaries before parliament. He struck the deals and issued the threats that Bute found so disagreeable, but which were so necessary for the smooth passage of the peace settlement. Fox did not rely solely on the powers of bribery and corruption so vividly described by Walpole, and often repeated in studies of these years.[75] In reality there was no

[73] *Bedford Correspondence*, iii. 141.

[74] *Grenville Papers*, ii. 2–3.

[75] Walpole, *Memoirs of George III*, i. 157. See for example, the repetitions of these ideas in: Rashed, *The Peace of Paris*, pp. 188–9 and Wiggin, *Faction of Cousins*, p. 278.

need for Fox to do much more than issue an appeal for loyalty to the king's wishes. In an atmosphere where the country was eager for peace and the terms negotiated were not discreditable, Fox had grounds for declaring that 'the peace will not long be unpopular, nor in Parliament at all so'.[76] The sabre-rattling of the opposition was an empty threat. The followers of Pitt and Newcastle had neither the organization nor a single common theme on which to unite a campaign against the preliminaries. To fulfil Temple's wild predictions of 150 MPs voting against the peace in the Commons would have required the sort of leadership that Pitt could at times inspire, but on this issue was sadly lacking. The great man was laid up with gout and no one had the slightest idea what he thought of the preliminaries.[77]

The final draft of the preliminaries was agreed and signed by ministers at a cabinet meeting on 10 November, with Grenville in attendance. Much to Grenville's mortification, certain issues over which he had fought with Bute were resolved as he originally desired. Compensation for Havannah was insisted on, for example, and came in the form of Florida. Even Bute's plan to lay the preliminaries before parliament for simple approval underwent modification. In an ingenious strategy the preliminaries were to be brought before the Commons after the king had ratified the terms under the great seal. This pledged the nation's faith in the settlement, making it impossible for parliament to break the peace or effectively alter its contents. Fox's confident handling of these preparations and his lack of concern at the threatened opposition proved fully justified as the preliminaries passed smoothly through parliament in early December. There were angy scenes at the opening of the new session on 25 November, as the mob vented its frustration on Bute.[78] Yet Fox was not mistaken in his belief that this mood of dissatisfaction would soon pass. This became evident when the preliminaries were laid before the Commons on 1 December, with a motion to approve them in both Houses eight days later. Nicholson Calvert for the opposition moved that discussion be adjourned for two weeks to allow Pitt the opportunity

[76] Ilchester, *Henry Fox, First Lord Holland*, ii. 210–11.
[77] *Grenville Papers*, ii. 3–7.
[78] *Grenville Papers*, i. 450.

to speak on the terms. The government refused the request, and, when put to the vote, carried the original motion by 213 to 74 votes.[79] This exploded the idea that the preliminaries would encounter any trouble from the newly formed opposition led by Newcastle's followers. On 9 December, when the government moved in both Houses for an address of thanks to the king thanking him for securing an end to hostilities, Fox's victory appeared complete. In the Lords one or two faint voices were raised against the terms, though not of sufficient force to warrant a division. In the Commons a similar fate looked likely for Newcastle's friends until Pitt made a late dramatic entry to the House swathed in bandages, intent on saying his piece about the terms. His speech lasted for three and a half hours, but it did not represent the damning oration that many expected. He criticized several points in detail, relating in particular to the articles on compensation and the Spanish fisheries. The most noteworthy clauses of his speech, however, were those castigating ministers for deserting Prussia and terminating Frederick's subsidy. This had been Grenville's pet scheme in the winter of 1761–62, and Pitt's attack emphasized the gulf that still existed between them. It also indicated why Grenville felt obliged to remain in office after his removal from the northern department, and he could not have regretted the decision on this day. Despite Pitt's theatrical performance, the government carried its address of thanks by 319 to 65 votes.[80] The following day ministers completed the rout of Newcastle's forces by defeating an opposition amendment to the address of thanks by 227 to 63 votes.[81]

Grenville must have experienced mixed feelings as he watched these events from the ministerial benches. The personal satisfaction he felt at Pitt's defeat had to be tempered with the knowledge that the detested Fox had engineered and executed the whole plan that carried the preliminaries safely through parliament. Grenville had just cause to express bitterness at Bute's refusal to grant him the powers enjoyed by every other leader of the Commons, especially in view of the fact that Fox managed the House

[79] Walpole, *Memoirs of George III*, i. 175.
[80] *Commons Journals*, xxix. 394.
[81] Ibid., p. 395.

without assuming a senior post. This underhand deal, as Walpole noted, allowed Fox to retain his lucrative post as paymaster while excusing him from any direct responsibility for the peace; 'Mr. Fox is again manager of the House of Commons, remaining Paymaster and waiving the seals; that is, will defend the treaty, not sign it. This wants no comment.'[82] Only once in the six months that he spent as first lord of the admiralty did Grenville publicly taste revenge. On 23 March 1763 Fox presented a petition for relief to the House from the inhabitants of Newfoundland who had suffered at the hands of the French the previous year. Fox's motives were humane enough, but he prepared the case badly. Grenville took a leading part in dismantling Fox's arguments, treating the petition with the utmost hostility. The points at issue were technical errors on Fox's part. First, he had made a case for relief that could not be substantiated by precedent. If there was no general indemnification for all sufferers in the war how could this partial claim be granted? Second, it had escaped Fox's notice that all the signatures on the petition were in the same hand.[83] His explanation that a clerk had copied out the original petition and had not had the time to have the copy re-signed was not accepted by the House. Embarrassed and outwitted, Fox withdrew the petition; 'he stood so ill in the public eye, that whatever he did received the worst construction'.[84]

This was small beer however in the light of Fox's success with the preliminaries, and the punishment of Newcastle's followers that followed. All those in office who had opposed the peace, and their dependents too, felt the sword of retribution wielded by Fox. In what became known as the 'Massacre of the Pelhamite Innocents', office-holders, important and insignificant alike, were removed and replaced by some friends of the government, whose loyalty to Bute and the king was not in doubt. The callous efficiency with which Fox carried this through put Bute's ministry in an impregnable position by the opening of the new session in January 1763. By this time even Newcastle and Pitt saw

[82] *Letters*, v. 26.
[83] *Commons Journals*, xxix. 603–4.
[84] Walpole, *Memoirs of George III*, i. 199.

the necessity of combining their efforts at Westminster to counter Bute's dominance of the Court, though they found it painfully difficult to agree on an issue with which to launch the first attack.[85] Indeed, the true measure of Fox's achievement became apparent in March 1763 when the government ran into difficulties over its proposal to levy a 4*s*. duty on cider but escaped without damaging its support. The tax was the brainchild of Dashwood, the chancellor of the exchequer. He introduced the proposed duty in the budget during early March, and it immediately aroused protest both inside and outside parliament. Much of the opposition bore a resemblance to the campaign against the excise scheme of 1733. In the vanguard of the protest against the cider tax were MPs from the producing counties in the West country who orchestrated an appealing campaign against the intrusion of the excise man into people's homes. Their cause was taken up by Bute's enemies in the City, who soon turned the protest into a struggle for personal liberty against the encroachments of the state. The articles written and published by John Wilkes in *The North Briton* and the speeches of Sir William Baker certainly declaimed violently against the threat to individual freedom but, at root, the campaign in London represented an ill-disguised tirade of abuse directed at Bute.[86] Be that as it may, the government had to defend the bill through every stage of its passage, and suffer a reduction in its majority. On 21 March, for example, the opposition moved to amend discretionary powers granted in the bill and met defeat by only eighty-three votes.[87] No real danger existed for the government. There was heat surrounding the issue, but Fox remained calm; he knew that Newcastle himself actually favoured the tax and found it difficult to commit himself whole-heartedly to the campaign. Yet Bute himself became unnerved by the resistance to the tax and decided to take steps to ensure its smooth passage. Shortly before the final reading in the Lords on 28 March, Bute met leading figures in the protest

[85] The behaviour of the opposition in these weeks is discussed in detail by O'Gorman, *The Rise of Party*, pp. 50–66.

[86] See, for example, the extracts from *The North Briton* cited in Nobbe, *The North Briton*, pp. 141–59.

[87] *Commons Journals*, xxix. 600.

campaign based in London in an effort to put an end to the protest. A deal was struck, the terms of which Bute revealed to the Lords during the final debate on the tax,

> Lord Bute said in the House of Lords that if the inconveniences should happen in the execution of this Act which are apprehended, that in that case he would himself move for the Repeal of it in the next session of Parliament.[88]

Henry Shiffner (MP for Minehead), who wrote this report on 31 March and who took a keen interest in defending the position of his constituents, welcomed this news but found it baffling. The concession was unnecessary in the strict sense that the government enjoyed large majorities during its passage; all Bute succeeded in doing was to convince people the government had no faith in the legislation. In that case, as Shiffner asked, why 'enforce a law liable to such uneasiness as I own I apprehend it may occasion?' If he had looked closer, maybe Shiffner would have seen in this the first public sign that Bute had had enough of battles at Westminster and the abuse that went with them. He had taken the decision to resign on 10 March, and was merely preparing the way for his departure.

Grenville played a miserable part in these proceedings. During the committee stages of the bill in the second week of March Pitt made a rare appearance in the House to harangue the authors of the cider tax. In a well-worn populist tone he re-enacted the struggle of 1733, declaiming 'against the dangerous precedent of admitting the officers of excise into private houses'. If this tax was revived, he said, it will necessarily lead to introducing the laws of excise into the domestic concerns of every private family'.[89] Grenville found such exaggerated rhetoric intolerable when the government had included clauses in the bill to guard against such abuses. He rose immediately and, in answer to Pitt, firmly defended the measure on the grounds of economic necessity. In the first place, cider, unlike beer, had never been taxed and here was an excellent opportunity to rationalize that anomaly to the country's benefit. Second,

[88] Dunster Castle MSS DD/L 2/44/11.

[89] This and the quotation that follow are taken from the account in *Parliamentary History*, xv. 1307–8.

though the excise was 'odious', no other traditional means of supply could engender the same revenue without considerable hardship. Feeling he had the upper hand, Grenville could not help trying to score another point. It was easy to be critical, he told MPs, but where would Pitt suggest another tax be laid? To emphasize the point Grenville repeated the phrase 'tell me where you can lay another tax' at least three times with increasing volume. At the finale of this crescendo a faint voice could be heard from the opposition benches singing the hymn 'Gentle Shepherd, tell me where'. It was Pitt's brilliantly sarcastic reply to Grenville's *tour de force*, and 'The Whole House burst out in a fit of laughter which continued for some minutes.' This was a savage blow to Grenville's confidence, and to make matters worse the name of gentle shepherd remained with him in newspaper articles and cartoons until his dying day.

In view of this wretched experience Grenville could hardly have been prepared for the letter he received from Bute on 25 March inviting him to take over as first lord of treasury.[90] Bute had openly talked for some time of his intention to retire once the peace settlement had been approved. This was achieved on 18 March when the final treaty was laid before parliament, and it seems that no amount of pressure from the king could dissuade him from this course. Though he knew his health was failing, Grenville only realized the seriousness of Bute's intention in the third week of March. Once Bute had convinced George III that he must let him go, the question of a successor immediately came up. Grenville, at that point, was the furthest person from the king's mind. As he told Bute on 14 March, 'Grenville has thrown away the game he had two years ago'.[91] Their preference not surprisingly fell on Fox who had carried the royal colours with such aplomb over the winter. Much to their amazement, however, Fox declined the opportunity to assume the rôle of leader. Like Bute, he wanted to retire to calmer pastures in the House of Lords now the task on the preliminaries was complete. His health would not allow him to take such an active position, he told the king on

[90] *Grenville Papers*, ii. 32–3.
[91] Sedgwick, *Letters from George III to Bute*, pp. 200–1.

17 March, and very reluctantly he recommended they turn to Grenville as the least objectionable alternative.[92] The king only came round to accepting this advice on 24 March, after Bute had done a considerable amount of preparatory work. Between the seventeenth and the twenty-fifth of the month he and Grenville held several meetings about the composition of a new administration, with their mutual friend, Gilbert Elliot, acting as an intermediary. It was not all plain sailing, as Grenville's head was spinning at the suddenness of this approach. He 'stated many difficulties in regard to his undertaking the situation proposed to him, but desired to be fully apprised of the whole before he could given any answer upon it.'[93] By 24 March sufficient assurances had been given to persuade Grenville to accept the post, bringing to an end the most remarkable month in his political career. Just as his morale reached its nadir he found himself being catapulted into the highest office of state, with Bute telling him that 'I still continue to wish for you preferable to other arrangements'.[94]

What were the preconditions that Bute apprised Grenville of to gain his acceptance of the changes? Unfortunately nothing in detail is known of the meetings between the two before 25 March. Judgements and assessments that have been made rely on the results of the talks rather than their content. Without exception they have all been damning to Grenville. Sir Lewis Namier believed that Grenville's ambition simply got the better of him, for he played no real part in forming his own ministry. Bute only consulted him after deciding who should take the senior posts in the administration, and that even Grenville's own treasury board was chosen in this way.[95] John Brooke supports this view of the obsequious Grenville swallowing all that Bute fed him: 'The Grenville ministry was intended as a mere façade. Bute was to remain the power behind the throne and final authority would rest with him.'[96] The basis of truth in this representation of the transfer of power from Bute to

[92] Ilchester, *Henry Fox, First Lord Holland*, ii. 231–2.
[93] *Grenville Papers*, i. 453.
[94] *Grenville Papers*, ii. 32.
[95] *House of Commons*, ii. 540.
[96] *King George III*, p. 103.

Grenville is borne out by the events of March and early April. Certain details in the overall picture do require revision, however, for these accounts give the mistaken impression that everything was cut and dried before Grenville became first lord of the treasury and chancellor of the exchequer on 13 April.

Grenville did not behave as the favourite's dupe as readily as it first appears. Though Bute's offer came as a surprise, Grenville was no longer a novice in negotiations of this sort. The experiences of the previous autumn were fresh in his mind as he entered the week of discussions with Bute, and he used them to good effect. They discussed several thorny problems about the new administration and on one of his issues Grenville succeeded in having his own way. Whether he intended to remain the power behind the throne or not, Bute made the most important concession of all by granting Grenville full powers of management and support in the Commons. The king expressed unease at this, but Bute impressed upon him the impossibility of Grenville succeeding as first minister without them. Furthermore, Bute took the trouble to tell all his associates that, as the king's choice, Grenville should receive their full support. 'I have filled my office with my friend G. Grenville', he told a Scottish associate on 9 April, 'whose integrity, ability, and firmness, I will be answerable for.'[97] On the question of places in the administration, it is also wrong to assume that Grenville played no part in the decisions taken before 13 April. On the contrary he scored two major successes. Prior to Bute's announcement of his intention to retire, Grenville's brother-in-law, Egremont, had also been threatening to resign over Bedford's conduct of the negotiations in Paris. Bute and the king wanted this to happen so that their protégé, Shelburne, could take the seals when Grenville came in. Grenville opposed this arrangement, and persuaded Bute to reconsider Shelburne for the post of president of the board of trade. Here, he argued, Shelburne's inexperience would not be exposed and his talents could flourish in the important job of organizing the government of the newly expanded American colonies.[98] Bute accepted this advice and, more

[97] *Caldwell Papers*, II. (i), 176.
[98] *Grenville Papers*, ii. 33–40 and Oswald, *Memorials*, pp. 413–17.

surprisingly, Grenville's recommendation that Egremont remain in place. Grenville had just cause to be delighted with this news. In the new administration he would be able to combine his efforts with the man he had so dominated during the previous year. The other secretary would be Halifax, who held no terrors for Grenville either. He did not have any say in this choice, but they had worked in harmony before and would do so again. A similarly revealing picture emerges when appointments in the new ministry are examined. Of two men at the treasury board, for example, which Namier pinpoints, Grenville expressed no reservations. Both Thomas Orby Hunter and James Harris knew Grenville and readily agreed to serve on his board. The same could be said of James Oswald who was promoted to the post of vice-treasurer of Ireland; his friendship and political connection with Grenville was of some twenty years' standing. Furthermore, Lord North who was already at the treasury welcomed Grenville's appointment whole-heartedly, as he had not found Bute a likeable superior at all.[99] Last, it is not strictly true to say that Grenville was denied a hand in appointments. Two of his recommendations, Thomas Pitt and Lord Howe, gained promotion to the admiralty board. None of this is really surprising in view of the fact that nearly all the appointments to the new administration, Grenville included, were members of Bute's political circle. Most of them had been connected with Bute since the Leicester House days, and they still looked to him for a chance of advancement. Indeed, if Grenville had been allowed a completely free hand in appointments to the administration he may well have come up with the same personnel.

Thus the traditional picture of Grenville assuming office as a phantom minister, surrounded by hostile colleagues, is hardly accurate. There were many imponderables that cast a shadow over the survival of Grenville's ministry. Would the king's determination to rely on Bute undermine confidence in the government? Would the inexperience of the new appointees like Shelburne and Lord Granby, master of the ordnance, jeopordize the success of Grenville's policies in parliament? Even Grenville's own qualities as outright leader

[99] Thomas, *Lord North*, p. 11.

had yet to be tested to the full. Nevertheless the future held some promise. The wily old Newcastle, observing these changes, was one of the few politicians to see the situation in a clearer perspective. People were right to laugh at Bute's attempts to create a proxy administration, he told Pitt on 9 April, but Grenville's qualities in office had to be taken seriously. If Bute wanted a puppet leader, he remarked, 'I question whether he has chosen the right person to act under him.'[100] This was the judgement of a man who had cast his mind back to the winter of 1761–2, when both he and Pitt had been outmanœuvred by Grenville. In these months Grenville had shown the cool, calculating streak necessary to lead the government successfully. Given the right conditions and support there was no reason why this exercise could not be repeated on a grander scale.

[100] *Chatham Correspondence*, ii. 222.

Chapter V

The Baptism of Fire: the first year as leading minister, 1763–1764

When the news of Grenville's appointment as the king's first minister became public in mid-April 1763, few contemporaries believed that the arrangement would last long. The shock of Bute's resignation prompted the view that any successor would suffer, most of all a politician who appeared to lack sufficient stature for the post. Bute 'is packing up a sort of ministerial legacy', wrote Walpole, voicing the general impression of the change in ministers, 'which cannot hold even till next session'.[1] Grenville himself seemed uncertain of the future. On taking up his appointment he made two unprecedented demands of the king: a pension of £3,000 per annum for himself when out of office and the reversion of a teller's place at the exchequer for his son. This had all the appearance of a desperate man exacting favours while the sovereign was temporarily at his mercy. It certainly offended the king to agree to these demands, who commented that Grenville's 'avarice overcame his prudence', and did not augur well for the future.[2] Indeed, the next twelve months did prove to be an interminable struggle for survival, but it did not end in disaster. On the contrary, Grenville confounded his critics by showing a degree of resilience and grim determination to succeed more than adequate to overcome the doubts and uncertainties of these early days in office. By the summer of 1764, Grenville's powers of leadership had not only swept the administration through a minefield of political crises but also established the strongest ministry in the first decade of George III's reign.

How did Grenville endure this turmoil? In the first few weeks of office it amounted to a case of ignorance being bliss. Grenville never perceived the handover of power from

[1] *Letters*, v. 301.
[2] Sedgwick, *Letters from George III to Bute*, pp. 230–1.

Bute in just the same way as his contemporaries or subsequent students of the period. Three overriding concerns guided his conduct in April 1763: avoiding isolation within the government's inner councils; securing the king's confidence, and, linked to this, ensuring that he obtained the full powers of patronage to manage the Commons. On the first point he felt reasonably confident. He had nothing to fear from the continuance in office of the two secretaries of state, Egremont and Halifax. On 4 April Bute had told Grenville to establish a 'strict union' with his secretaries 'as the only means of supporting the king's independency', and the advice was followed to the letter.[3] The triumvirate, as Grenville, Egremont, and Halifax soon became known, determined from the beginning of the ministry to exercise personal control over all important government business, becoming in essence a three man cabinet. Sir Lewis Namier says this was an 'anamolous formation', but, in view of the uncertain atmosphere in which Grenville came to power, a formation for self-preservation would seem a better description.[4] To keep decision-making on the most important issues within this small group lessened the risk of internal disagreements, and, more important, reduced the chances of seeds of disunion being sown by the ministry's opponents, who were preparing to attack even before Grenville took office. The novel part of this arrangement came with Grenville's decision to allow Egremont and Halifax a share of patronage in the new administration. This action is frequently portrayed as evidence of weakness at the core of the new ministry, for no other first minister had made such a concession. In general there is substance to this claim, because Grenville of the three ministers in the triumvirate suffered most doubts about the future after Bute's sudden withdrawal from public office. Yet in practice sharing patronage did not prove the self-incriminating sacrifice that is often cited. Such was Grenville's influence over Egremont, his brother-in-law, that no difficulties over appointments ever arose, even in the event of a dual recommendation. Grenville's willingness to make this concession to Halifax, also sealed

[3] *Grenville Papers*, ii. 40.
[4] *House of Commons*, ii. 540.

the bond of trust that existed between the ministers in the triumvirate. Halifax could have been difficult about serving under Grenville. He was a politician of higher rank and status than Grenville, and it behoved the new first lord to gain Halifax's confidence as soon as possible. This gesture achieved that end, and the system of conducting business in this closed circle served its purpose well until Egremont's unexpected death in late August 1763.

On the second and third points Grenville was at first content. The king had agreed in late March to let Grenville have full powers to manage the Commons without Bute's intercession. The favourite assured Grenville that 'he was determined to be a private man for the rest of his days, never to intermeddle in Government, and that he was going out of town with his family to drink the waters at Harrogate.'[5] From the evidence available there is no reason to question Bute's sincerity on this point. At the end of April he did go off to Harrogate with his family for a holiday, and then retired to his country seat at Luton Hoo for the summer. The king's confidence in Grenville was a more problematical issue. If Grenville made one grave error in these early weeks of office it was to accept as the gospel truth Bute's assurances of the king's commitment to the new administration. Bute may have told his followers to transfer their allegiance to Grenville but it soon became apparent that the king had no intention of doing so. No one could have predicted how stubborn and childish the king would be as he sulked over the loss of his 'dear friend'. It was George III and not Bute who paid lip-service to the promises of support given to Grenville before he took office. In private the king made it clear that he intended to consult the favourite before he approved any of Grenville's policy decisions. Furthermore, he insisted after Grenville became first lord that Scottish patronage remain in Bute's hands, and even hinted that ecclesiastical appointments should come under his control too. The letters sent by George III to Bute in the spring of 1763 show that the king took out much of his frustration at Bute's retirement on the hapless Grenville.[6] The smallest

[5] *Grenville Papers*, i. 453.

[6] Sedgwick, *Letters from George III to Bute*, p. 231 et seq.

slip by Grenville, such as the time in April when he forgot to make an appointment for Sir Francis Dashwood to kiss hands for his peerage, provoked angry outbursts from the king at the 'neglect' of his first lord. Sulking it may have been, but it undermined the basis of trust between king and first minister so vital to Grenville's task of governing successfully. If the king's attitude remained unchanged Grenville's position would become untenable. John Brooke did not exaggerate when he wrote 'If ever King George III can be accused of behaving in an unconstitutional manner it is during the first few months of the Grenville ministry.'[7]

Grenville's suspicions about Bute's continuing influence were first aroused when he drew up the king's speech at the close of the session on 19 April. George III would only agree to the wording after Bute had approved the final draft. For the rest of the month the king bombarded the favourite with pleas for help in taking decisions and requests to stay in London. Grenville tolerated this behaviour in expectation of Bute's departure for Harrogate on 2 May, for at least with the favourite out of the city it would be physically impossible for the king to consult him. Yet George III remained moody and distant with all his senior ministers, casting grave doubts about their future tenure in office. He took a particular dislike to Grenville's self-righteous and verbose manner in the Closet. The king complained in his letters to Bute of Grenville's 'tiresome manner' and 'selfish disposition', and it became clear from an early date that the king's personal feelings about Grenville would bedevil the ministry's chances of survival. Historians have made much of this enmity between king and first minister, with the blame usually being laid squarely at Grenville's door. They found no excuse for the undignified way Grenville demanded favours of the king and tried to intrude himself on George III's private business, dictating appointments to offices about his person. Lord Holland's comment of August 1765 about Grenville's relations with the king is the one side of the argument always presented 'I am persuaded, Selwyn, that the king, who we can see can swallow anything almost, could not, however bear his [Grenville's] conversation. A dose so large and so

[7] *King George III*, p. 103.

nauseous, often repeated, was too much, for any body's stomach.'[8] Such biased comments have been taken too far, however, perpetrating the impression that Grenville was insensitive and callous in all his personal relationships. This was certainly not the case as his family, friends, and followers testified during and after his life. True, Grenville could be wearisome in conversation and prolix in the Commons chamber: he did not suffer fools gladly either. Those who met him usually came away attracted or repelled, few expressed indifference. But his relations with the king were a unique problem. On official business George III and Grenville saw eye to eye. Two problems dominated Grenville's ministry: John Wilkes and American colonial policy and no amount of pressure or criticism shook their unity on these issues. Yet the king's personal antipathy never lessened, and Grenville never really understood why. If he had read his wife's diary perhaps he may have found a clue to the mystery. The overriding feeling evident in the record of these uncertain months is that Grenville simply did not trust George III, and would not take his word for granted. In view of the way he had been treated over the previous year he can hardly be blamed for this. Indeed, as the events of the summer of 1763 unfolded it would have been folly to adopt any other attitude, as the king continued to mislead and deceive.

At first the king attempted to be subtle in his wish to change the complexion of the administration. When Bute returned to London on 1 June the king authorized him to make discreet enquiries of the opposition leaders, to see whether or not they would be willing to serve individually or come in with some of their followers. Over the next six weeks Bute contacted Lord Rockingham, the Duke of Newcastle, and the Earl of Hardwicke, receiving a negative reply from all three. The king excused these approaches to Grenville by saying that he looked to strengthen the ministry by filling the post of lord president, vacant since Granville's death in the new year. Grenville knew otherwise, however, for in his wife's diary the approaches to Hardwicke are described in detail, and they concerned a good deal more

[8] Jesse, ed., *George Selwyn and his Contemporaries*, i. 405–6.

than simply tampering with one or two offices: Hardwicke would only serve 'upon a plan concerted with Mr. Pitt and the great Whig Lords, as had been practised in the late king's time'.[9] It did not take Grenville long to see the hand of Bute, or, as he called him, 'that superior influence', behind these moves, and in concert with Egremont and Halifax he decided to lay the matter before the king with a demand for an end to the double-dealing. In fact the two secretaries were for resigning immediately if they did not receive firm assurances of support from the king. Grenville talked them out of this by convincing them that resignation would be playing into the hands of their opponents. Pitt was the only alternative leader at this juncture, and Grenville wanted the king to take the intiative on bringing him in, as opposed to a protest resignation which would allow George III the satisfaction of saying he had been deserted and left to the mercy of his enemies. This confident response was borne out of some hard experience since April 1763, and a realization that this fight for survival had to be won if the prize of a lifetime's labours was not to be thrown away. Initially Grenville succeeded in this strategy of calling the king's bluff. George III responded to the triumvirate's ultimatum by asking for ten days to consider matters. In the interim Bute approached Bedford through Shelburne to see if he would go where others feared to tread. Bedford was more responsive than the others, laying down two conditions for entry to office: first, Bute must retire completely from politics and second, Pitt had to take the lead in the Commons. All hinged on the response of Pitt, but, when asked about Bedford's plan, he said that he would not serve with him, or for that matter any one connected with settling the terms of the Peace of Paris. The king appeared defeated, and on 21 August Grenville was summoned to his presence to be told

> that he had no desire to change his Ministers; on the contrary he liked them all, he approved of their conduct, and meant to strengthen them by every means in his power; that he had sent for Mr. Grenville to

[9] This and the quotation that follow are taken from Mrs Grenville's diary covering the events of July and August 1763, *Grenville Papers*, ii. 190–3. They are discussed in more detail by Tomlinson, M.A. thesis, pp. 17–47.

talk with him about it, to enter fully into the matter, and to desire him to bring him his thoughts in writing upon such measures as he should think expedient . . . for that by his advice he meant to conduct himself.

This audience of the king marked a watershed in Grenville's relations with the sovereign. If matters had remained unchanged, he might just have been able to forgive and forget the backstairs manipulations in June and July, and accept the king's declaration of faith in good heart. What happened next ruined any hope of this. On his way home from this interview on 21 August Grenville called on Egremont to tell him of the outcome of their strategy, where he learned to his shock and horror that the secretary had been seized with an apopleptic fit and died. All the anguish of early August now seemed wasted, for Grenville immediately saw that the king might use Egremont's death as an excuse to replace his leading ministers. He was not mistaken, but, in this case, Grenville did not wait for a royal sign that his days in office were numbered. On 23 August he met Halifax, and they decided to take the initiative and present the king with another ultimatum with regard to Egremont's demise, stating 'that his Majesty had three options, either to strengthen the hands of his present Ministers, or to mingle them with a coalition from the other party, or to throw the government entirely into the hands of Mr. Pitt'.[10]

There is little doubt from Grenville's account of this audience that he and Halifax believed that they had secured another victory. Of the three alternatives the king discussed, Pitt's coming in was 'what he never could consent to'. This left what Grenville considered to be the only sensible course, an accession of new blood to strengthen the existing administration. How wrong could he and Halifax have been? The next day Grenville learnt that the king's statement was a downright lie. Bute saw Pitt in person on 24 August, and, according to Grenville, 'settled with him the terms upon which he and his party would come in'. At this point, feeling utterly betrayed, Grenville lost all faith in what both Bute and the king said. When questioned, they would not even

[10] *Grenville Papers*, ii. 194–5. The account and quotations that follow are also taken from here.

own up to the approach: 'Lord Bute and his friends deny this', runs the sour note in Mrs Grenville's diary. During his attendance at Court on 25 August it is clear Grenville expected to be told something of the negotiation but 'nothing passed'. The thunderbolt was delivered the following day 'when the king opened to Mr. Grenville his intention of calling in Mr. Pitt to be management of his affairs declaring that he meant to do it as cheap as he could, and to make as few changes as was possible.'

At this low point in Grenville's short career as first lord his patience ran out. In response to the news of Pitt's return to favour, Grenville gave the king a blunt reminder of his previous declarations against such a move; adding for good measure that he 'saw no necessity in the king's affairs which could urge such a surrender, and that he could never be a party to it, it being impossible for him to act with people whose every principle was so diametrically opposite to his own.'[11] In the evening he gave the matter further thought and told Halifax that he had had enough. Though he knew what the answer would be, Grenville intended to force the issue with the king by asking him whether he should or should not go on with treasury business. Halifax sympathized with Grenville's attempt to salvage his self-respect from the situation but refused to countenance any threats of resignation. In a reversal of the rôle earlier that summer when Grenville put the brake on his two secretaries, Halifax advised caution. Such an enquiry, he told Grenville 'would look, I think, too much like asking your discharge, and bear the appearance of ill humour'. Perhaps Halifax had an idea the negotiations with Pitt might founder and the game could yet turn in their favour, for this advice proved well founded. More important, Grenville fortunately saw fit to accept it. When he saw the king on 27 August his purpose was not to bring matters to a head but to launch into another tirade on his sufferings over the previous five months. The atmosphere was strained for George III had just been with Pitt for two hours, and, contrary to all expectation, failed to reach agreement over his terms for entry to office. In short, Pitt wanted wholesale changes in personnel and policies, reflecting

[11] Ibid., p. 196.

his desire for a return to the wartime ministry with Newcastle and himself at the helm.[12] Not surprisingly the king baulked at resurrecting a system of government which he had been dedicated to overthrowing since his accession, and their talks ended in deadlock. On entering the Closet after Pitt, Grenville sensed all was not well. The king appeared 'a good deal confused and flustered', and Grenville thus decided to press home the points made the previous day. In the circumstances it was not a bad idea to remind the king of his obligations to ministers who were still loyal and willing to serve. In characteristic style, however, Grenville overdid the sermonizing. Exasperated at a day of fruitless negotiation and Pitt's audacity, the last thing the king needed was a lecture on the morality of his behaviour in the preceeding weeks. He listened for twenty minutes, then could take no more. George III 'bowed to Mr. Grenville told him it was late, and, as he was going out of the room, said with emotion, "*Good morrow, Mr. Grenville*", and repeated it again a second time, which was a phrase he had never used to him before.'[13]

If Grenville failed to profit from the king's premature enthusiasm for Pitt's return to office, he would only have himself to blame. A conciliatory word could have made all the difference to the king's mood on the twenty-seventh, but Grenville could not find it in him to utter one. So suspicious was he of the king's actions by this point, that outright aggression governed all his conversations with the sovereign. This defensive mechanism was Grenville's method of survival and it almost brought him down. He was rescued in part by the devotion of his friends Gilbert Elliot and Charles Jenkinson, who talked with Bute on the morning of 28 August and 'terrified him so much upon the consequences of the step he had persuaded the king to take, that he determined to part from it, and to advise His Majesty to send to Mr. Grenville'.[14] Bute did remain true to his word, and his advice certainly swayed the king's decision to reject Pitt's terms. In the main, however, it was by a stroke of luck, and no more, that the king deemed Grenville's lectures preferable

[12] For a full account of these negotiations see Hardy, M.A. thesis, pp. 122–39.

[13] *Grenville Papers*, ii. 196.

[14] This and the quotations that follow are also taken from the *Grenville Papers*, ii. 197–201.

to Pitt's bombast and extravagant demands. If a little modesty had prevailed in Pitt's dealings with the king, Grenville could have been out of office on 28 August. The king's personal dislike of Grenville was as strong as ever when he called him to an audience later that day and informed the long-suffering first lord

that he wished to put his affairs into his hands; that he gave him the fullest assurances of every support and every strength that he could give him towards the carrying his business into execution; that he meant to take his advice, and his alone, in everything: that it was necessary the direction should be in one man's hands only, and he meant it should be in his.

Both parties realized at this interview that a good working relationship was the best that could be hoped for in the future running of the government. The basis of trust and mutual understanding necessary for a deeper personal friendship had crumbled away over the summer months.

Knowledge of this fact did not upset Grenville too much. To survive, a first minister required the support of the Court and a mastery over the House of Commons. Grenville had triumphed on the first point and was confident and experienced enough to realize this victory must be consolidated before he could hope to establish a dominant position at Westminster. Grenville wasted no time in laying the groundwork for strengthening his hold over the king. After accepting the invitation to continue as first lord, Grenville insisted that the following conditions be observed: 'that to enable him to make that duty and zeal of real service to His Majesty, he must arm him with such powers as were necessary, and suffer no secret influence whatever to prevail against the advice of those to whom he trusted the management of his affairs.'

The wording of this entry in Mrs Grenville's diary is significant. In June Bute appeared as 'that superior influence'; by August he had become the mere sinister 'secret influence', and from this moment the myth of the minister behind the curtain took root in Grenville's brain, colouring all his dealings with the Court thereafter. Though he heartily disagreed with such terminology, the king had little option but to accept Grenville's terms; and on the following day, 29 August, showed him a letter from Bute confirming his retirement 'from reasons of nationality, unpopularity etc. etc.'.

For the moment Grenville took this declaration in good faith and fears of further double-dealing faded with Bute's departure from London.

With the favourite gone, the final link in completing this stranglehold on the king would be the choice of ministers to fill the vacant posts in the administration. The immediate priorities were the southern secretary's post held by Egremont until his sudden death, and Granville's old office of lord president of the council. The former posed no difficulties for Grenville and the king. Halifax moved from the northern to the southern department and Sandwich stepped up from the admiralty to fill that vacancy. The presidency of the council proved a more thorny problem. Grenville's inclination, not unnaturally, was to favour a compliant candidate, and on 5 September he offered the job to the ailing Duke of Leeds. To his dismay, the old duke refused on grounds of ill-health, prompting three days of intense deliberation between Grenville and George III over who best to approach next. The choice eventually fell on the Duke of Bedford. But it took Grenville the whole three days to convince himself that this was the right step. In view of his behaviour during the recent negotiations, could he be trusted to follow Grenville's lead? He was a powerful figure, who could ensure an increase in the government majority of approximately twenty votes.[15] The problem, as Grenville explained to the king on 8 September, was that he might prove too powerful:

> [the king] had advised him to call to his Government, not such as were his friends, viz. the Duke of Bedford and Lord Sandwich, but such as he thought could best strengthen it; that these might prove too strong for him, his only reliance was upon His Majesty's truth and honour, and on that he trusted he might depend.[16]

The king dispelled these fears at once, assuring Grenville that he could depend upon his 'trust and honour . . . that he would never fail him, nor forget his services'. The king proved reliable on this point. Bedford and Sandwich assumed their new offices under the impression that the system of sharing patronage, existing under the triumvirate, would be revived.

[15] There is a description of Bedford's parliamentary following in Brooke, *The Chatham Administration*, pp. 255–62.

[16] *Grenville Papers*, ii. 205. The account and quotations that follow are taken from Mrs Grenville's diary too pp. 205–7.

Grenville was too wise now, however, to allow that. If he gave in on this issue at that moment, he would have no chance of establishing his mastery over the Commons. This could only be achieved by the creation of a body of followers at Westminster devoted to Grenville's cause, and answerable solely to him. The king understood the patronage situation perfectly and rebuffed private approaches from both Bedford and Sandwich on the question of sharing. As he told Grenville on 15 September, 'he meant to take Mr. Grenville's recommendations and his alone . . . for that he meant that all graces should be done through Mr. Grenville alone'. From this point on, Grenville was minister in fact as well as name.

It would be a source of eternal regret for Grenville that having survived these machinations of the Court, his personal relations with the king did not improve. There were perennial disputes on patronage and favours throughout Grenville's ministry that simply deepened the uncertainties each felt about the other's intentions and loyalties. In some of these disputes, like the request in April 1764 for apartments overlooking the king's private gardens at Richmond, Grenville is shown in a poor light.[17] He is either offending George III by deliberately claiming the right to supervise appointments about the king's person or gratuitously seeking favours for his own family in lieu of services rendered at Westminster. In other arguments the blame is not so easily apportioned, for they usually arose from misunderstandings on both sides. The problem was epitomized in September 1763 by the struggle to reappoint a successor to Bute's old sinecure post as keeper of the privy purse. The office had never been considered a political appointment, and the king took umbrage when Grenville suggested that Lord Guilford, and not the king's recommendation Sir Willian Breton, be given the post. Grenville's motives in this have been ascribed to a desire to enlarge his following in the Commons — Guilford's son was Lord North.[18] But there was more to it than that. Breton was a close friend of Bute, and Grenville saw in his

[17] Sedgwick, *Letters from George III to Bute*, p. 237.

[18] *King George III*, p. 106. It is possible Grenville chose Guilford in particular because he had been the king's governor before the death of Frederick, Prince of Wales, and George III seems from later evidence to have had respect and regard for him. Grenville might have thought him on that account an acceptable alternative to Breton.

appointment an immediate link with the favourite, and an attempt to recreate by proxy 'that secret influence' over the king. Furthermore, Gilbert Elliot had told Grenville on 31 August that Bute himself believed the king would consult the first lord over the vacancy, adding for good measure that 'He seems extremely desirous that you may be armed with every degree of power necessary at this juncture.'[19] It is for these reasons that Grenville battled with George III over the appointment, and, even when defeated, continued to tax the king about Breton's position. Eventually the king lost his temper with this never-ending enquiry, exclaiming 'Good God! Mr. Grenville, am I to be suspected "after all I have done?".' It is a sad truth that had he really wanted an answer the reply would have been in the affirmative.

The remarkable fact about this constant wrangling with the king is that Grenville still managed to secure his second goal of primacy at Westminster. For the next eighteen months the demands of official business overcame their personal antipathy. Grenville's triumph over the summer goes some way to explaining this. He now carried the prestige of being the king's minister, with sole control of patronage in the Commons, and this gave him a self-confidence that was soon reflected in his performance in the House. In this light the events of the summer may be seen as a blessing in disguise, for Grenville was required to assert his qualities of leadership far earlier than he anticipated. He had barely had time to draw his breath after defeating Bute before his ministry was rocked with a political crisis that threatened its survival. The point at issue was the *North Briton* No. 45, published on 23 April 1763, only ten days after Grenville became first lord. This issue of the political weekly contained an inflammatory attack on the king's speech, delivered on 19 April at the close of the parliamentary session. The most offensive passage charged George III with being 'brought to give the sanction of his name to the most odious measures, and to the most unjustifiable public doctrines, from a throne ever renowned for truth, honour, and unsullied virtue'. What had escaped as polemic in previous issues was now seen as a personal affront to the king's honour, and ministers decided

[19] *Grenville Papers*, ii. 210. The quotes that follow were also taken from here.

on prompt action to deal with the author. There would be no difficulty in this for it was widely known that John Wilkes, the opposition MP for Aylesbury, had penned the offensive piece. On 25 April Halifax referred the paper to the government's legal officers for an opinion, and attorney-general Charles Yorke and solicitor-general Sir Fletcher Norton advised them that in their view it constituted a seditious libel. On receiving this news Halifax employed the time-honoured executive mechanism for silencing newspaper critics: on 26 April he issued a general warrant for the arrest of the authors and printers of the *North Briton* No. 45. The arrests began on 29 April.[20]

To this point Grenville had no direct involvement in the case, for it was a straightforward departmental matter. Halifax guided the strategy, with his efficient under-secretary, Edward Weston, overseeing the paper work.[21] What happened next, however, would make Grenville rue the day he was not told of the affair in more detail. The first printers apprehended on the general warrant gave evidence to prove that Wilkes was the author of the offending article in the No. 45. On the strength of this testimony, Norton and Yorke gave their opinion that Wilkes himself could now be lawfully arrested. They believed the libel to be a breach of the peace, and thus not subject to a claim of immunity from prosecution on the grounds of parliamentary privilege. Halifax, in concert with Egremont, then proposed to issue a new warrant naming Wilkes, but P. C. Webb, solicitor to the treasury and Lovell Stanhope, law clerk to the secretaries of state, counselled that such a step would be unnecessary. They maintained that Wilkes could be taken into custody under the original general warrant, and won the approval of their superiors for carrying it into effect. The fatal legal blunder, which Grenville's opponents later dubbed an attack on the constitution, had been committed; though in truth the motives of Webb and Stanhope were no more sinister than wishing to appear knowledgeable and efficient. Wilkes was arrested on 30 April, and appeared in court on 6 May. His case against the warrant was based simply on parliamentary

[20] For further background on these events see Thomson, *The Secretaries of State*, pp. 112–26.

[21] H.M.C. 10th report, pt. 1, p. 355.

privilege, and in a sensational judgement Wilkes secured his release on these grounds. For this Wilkes had to thank his own legal advisers, who brought the appeal against his apprehension before the chief justice of the Court of Commons Pleas, Charles Pratt, and not in the King's Bench Court under Lord Mansfield. Indeed, throughout the debates on general warrants and Wilkes's arrest, lawyers sympathetic to Wilkes's cause proved a constant irritation to the government by exposing technical blunders in their case and exploiting legal loop-holes to the advantage of their client.[22] Pratt's decision of 6 May was the first of many that caused consternation in Halifax's department, because his servants knew that if Wilkes was to be punished now, parliamentary intervention could be unavoidable.

It was some weeks before ministers took decisive steps to rectify the errors of their law officers. The government did commence an action against Wilkes in the King's Bench Court, but as long as he enjoyed immunity from prosecution the matter could hardly be resolved. This indecision reflected Grenville's uncertain future over the early summer. If there were serious doubts about his chances of surviving as first lord, what point was there in planning ahead to the next session of parliament? Wilkes himself suffered no such inhibitions. Encouraged by popular support for his stand against general warrants, his presses began printing again right after the acquittal. In June the infamous *North Briton* No. 45 was reprinted and, to add insult to injury, Wilkes and the printers began legal proceedings against the government for damages arising from the original arrests. This provocation was sufficient to prompt both Halifax and the king to raise the matter with Grenville in late August.[23] But nothing was done in earnest until the following month when Grenville had secured his position as first lord. By this time the situation looked fraught with difficulties, as the opposition was threatening to use the issue to attack the ministry at Westminster.[24] And on this occasion they would

[22] The illuminating essay by Brewer, 'The Wilkites and the law, 1763–74', gives a full description of the work of Wilkes's legal advisers and explains its significance in *An Ungovernable People*, pp. 128–72.

[23] *Grenville Papers*, ii. 192 and Tomlinson, *Additional Grenville Papers*, p. 28.

[24] Hardy, M.A. thesis, pp. 138–51.

be able to count upon the valuable assistance of Pitt, who believed that the government's actions represented an attack on MPs' privileges and a threat to the freedom of the press. Grenville thus took control of preparing the parliamentary strategy himself, and in mid-September began collecting information on legal points relating to the case.[25] He was assisted by the new northern secretary, Sandwich, a former colleague of Wilkes. Their relationship had soured in the last days of the notorious Hell Fire Club, and he was now eager to see Wilkes punished for his effrontery. He and Grenville thrashed out the basic strategy by which the government would move against Wilkes at Westminster. Their plan was straightforward. A complaint had already been laid against Wilkes in the King's Bench Court and it was the ministry's intention to convince parliament that he should answer it. Initially, the No. 45 would be brought to the attention of the House, during the Address on 15 November, as a seditious libel with a recommendation to punish the author. If this was accepted by a large majority Grenville's action would then be *ad hominem*; to persuade MPs that Wilkes, the individual, ought not to be covered by privilege from criminal prosecution, and that in order to strip him of this protection he should be expelled the House.[26] Grenville prepared the government plan meticulously, for this time there would be no technical blunders. As he explained to his colleagues in October, there was no room for vendettas: the government was only concerned with the question of privilege 'Any censures, or even the voting of expulsion were not be considered as punishments, for then (supposing a prosecution elsewhere) a man for the same offence would be punished twice, . . . such acts of the House were to be considered *as vindicating of their own Honour*'.[27]

Grenville attached great importance to the success of this policy. In conversation with James Harris on 29 October he declared that 'if this question were lost he must immediately

[25] B.L. Add. MSS 57810, fo. 55.

[26] For further detail on the preparation of the case against Wilkes see Lawson, Ph.D. thesis, pp. 87–92.

[27] Memo of James Harris dated 19 Oct. 1763, cited in Tomlinson, M.A. thesis, pp. 57–8.

quit administration'.[28] He had assumed full responsibility for the attack on the *North Briton* and would stand or fall on its reception at Westminster. Grenville intended to show the political world that his ministry could survive without Bute: 'this would be a test', 'whether Government could support itself or not'. Preparations for the case dominated ministerial business from September to mid-November. Grenville did foresee some difficulties over the cider tax in the coming session, for Bute had promised to review the legislation in March 1764. With some judicious modification, however, Grenville felt confident that the rumblings of a campaign for repeal amongst the West country MPs would be nipped in the bud.[29] At this juncture there was no appreciation at all of the important rôle general warrants would play in the political battles ahead. Sandwich echoed Grenville's sentiments to the letter as he assisted in amassing the legal precedents for the expulsion plan. 'That spirit of licentiousness which has but too much manifested itself in this country', he told Gilbert Elliot in early November 'will be timely and effectively suppressed by the vigorous interposition of the authority of parliament against the authors and abettors of the scandalous and infamous libels that have been published here during the course of the last summer.'[30] This attitude contrasted sharply with Halifax's languid approach in the early summer, but it reflected the fact that Grenville and his new-found allies, the Bedfords, were now playing for much higher stakes. The urgency evident in these preparations was based on the practical necessity to execute the policy against Wilkes before he appeared in the Court of Common Pleas with his appeal for damages. The hearing had been set for 6 December, and, as Sandwich told Webb in late October, would certainly go Wilkes's way before a London jury and with Pratt presiding.[31] It was imperative in the government's view, therefore, that to avoid trouble in the City and at Westminster, Wilkes be made to answer the complaint in the King's Bench Court before 6 December.

[28] This and the quote that follows are taken from memos of James Harris also cited in Tomlinson, M.A. thesis, pp. 48–9.

[29] Ibid.

[30] Minto MSS 11036, fos. 61-2.

[31] Guildhall MSS 214/1, 203.

As MPs assembled for the opening day of the session on November 15, Grenville felt confident that his plan would succeed. In late September, he had told Bedfordite Richard Rigby that the opposition would pose no threat to the government's position on the issue. 'I hear from Grenville', Rigby reported to one ministerial supporter, Sir John Wynn, 'the best appearances of our majority next sessions, which promises to be very near if not quite as large as the last.'[32] Other contemporary analyses bear out Grenville's impression that the government had little to fear. J. R. G. Tomlinson's investigations have shown that two lists of prospective supporters and opponents were prepared before parliament met: one by Thomas Orby Hunter at the treasury and the other by Charles Jenkinson. They reached different conclusions, mostly through errors in classification of MPs, but neither doubted the government's majority would be over 150 votes. Tomlinson himself also made an excellent analysis of the 586 MPs who sat in the Commons during Grenville's ministry and found that both contemporary lists were not far off the mark, In his calculations 358 were government supporters and 182 in opposition.[33] Thus Grenville faced the House in good spirits on 15 November when he presented a message from the king requesting MPs take action to ensure that Wilkes answer the complaint in the King's Bench Court. Everything pointed to a successful outcome. Before the debate began, Grenville's message had been given priority over an appeal from Wilkes claiming a breach of privilege, after a vote of 300 to 111 had underlined the ministry's strength.[34] Hopes that the main question would be resolved in similar fashion were soon dashed, however, with Pitt's intervention in the debate. In a long and sometimes fierce discussion, freedom of the press and arbitrary arrest were debated at length, and it was one o'clock in the morning before the motion declaring the *North Briton* No. 45 'a false, scandalous and seditious libel . . .' could be put to the vote. In the end there were no surprises: in a discussion over the wording of the motion Grenville triumphed by 273 to 111 votes.[35]

[32] Glynllifon MSS 253 (30 Sept. 1763).
[33] M.A. thesis, pp. 105–10 and 657–706.
[34] *Parliamentary History*, xv. 1354–5.
[35] *Commons Journals*, xxix. 667–8.

The next stage of Grenville's strategy was completed on 24 November when Lord North presented the House with a resolution 'that privileges of Parliament does not extend to the case of writing and publishing, seditious libels'.[36] Again Pitt spoke against denying Wilkes his constitutional rights. The House had done its duty in defending the king's honour on 15 November, he told MPs: 'But having done this it was neither consistent with the honour and safety of parliament, nor with the rights and interests of the people, to go one step farther. The rest belonged to the courts below.' His argument had weight, but the case overall suffered from an appearance of sympathy with Wilkes the man, whose devotion to drunkenness and debauchery found few defenders in the House. In the division, North's motion gained a comfortable majority by 258 to 133 votes.[37] Stripped of his parliamentary immunity, Wilkes began to prepare for his flight to France rather than face the courts. In the second week of December he left Dover for Calais, and eventually made his way to Paris. All that remained for Grenville was to execute the plan for expulsion after the Christmas recess. The government moved against Wilkes on two points in a plan devised by Sandwich.[38] The centrepiece of the strategy was a proposal to condemn not only the No. 45 before the Lords but also Wilkes's obscene poem 'Essay on Woman' published privately in July 1763. To ensure that there would be no mistakes, Sandwich himself established the legal grounds for prosecution and the co-operation of the government's leading witness, Bishop Warburton. It was a crude arrangement, consisting of little more than a character assassination; yet, to Sandwich's credit, the whole was put into effect without a hitch. On 15 November the Lords discussed and condemned both the *North Briton* and the 'Essay on Woman', and two days later agreed to order that an Address be presented to the king for the prosecution of the author for 'the said scandalous and impious libel'.[39] After the Christmas recess the Commons followed suit and

[36] This and the quote that follows are taken from the account in *Parliamentary History*, xv. 1362–4.

[37] *Commons Journals*, xxix. 675.

[38] Guildhall MSS 214/1, 214.

[39] *Lords Journals*, xxx. 420–1.

Wilkes was ordered to attend the House on 19 January. In view of his flight to France, Grenville knew there was little chance of his making an appearance and he decided to push through his policy of expulsion in Wilkes's absence. On the nineteenth, Wilkes was found guilty of contempt of the House's order, and, in the early hours of the following day, was expelled for the offence of 'writing and publishing the *North Briton* No. 45'. There were four divisions on this day and the minority vote fell from a high of 102 votes to a low of 57 votes.[40]

Grenville's immediate reaction to these events was understandable relief. The king expressed 'his satisfaction', noted Mrs Grenville in her diary, and 'the whole was very triumphant'.[41] So much had been risked in the strategy to deal with the *North Briton*, and it succeeded without any threat to the ministry's parliamentary position. To add to his joy, Grenville followed this triumph with a conclusive settlement of the troublesome agitation over the cider tax. It did not prove easy at first. Bute's promise of a review for the previous year had encouraged petitions and, in some western counties, a good deal of rioting. In the new year the opposition took up the cause with Pitt's blessing, and lent its support to William Dowdeswell, Tory MP for Worcestershire, and recognized leader of the protests against the tax. Indeed, it was Dowdeswell who reopened the issue at Westminster by moving on 24 January 1764 for a Committee of the Whole House to debate the issue. In keeping with his views outlined to Harris in October, Grenville sought to restrict the powers of the Committee so that only alterations and not repeal of the tax would be considered. In a division on this proposal Grenville had his way but not by an overwhelming margin: the vote of 167 to 125 showed a residual dissatisfaction at the operation of the tax.[42] This point was underlined on 31 January when Dowdeswell forced another vote over his proposal that the tax should fall on the retailer rather than the manufacturer. The government only managed to defeat this motion by the slim margin of 172 to 152 votes.[43]

[40] *Commons Journals*, xxix. 722–3.

[41] *Grenville Papers*, ii. 484.

[42] *Commons Journals*, xxix. 737.

[43] This is taken from the diary of James Harris, 31 Jan. 1764, cited in Thomas, *British Politics and the Stamp Act Crisis*, p. 18.

At this juncture Grenville decided to intervene with his own proposals for rationalizing the burden of the tax, which found favour with the majority of MPs from the producing areas.[44] On 7 February the Committee discussed these alterations to the cider legislation and accepted them in principle, guaranteeing Grenville another parliamentary success. On 10 February an opposition motion for outright repeal of the tax was defeated by 204 to 115 votes.[45]

These proceedings should have been the last act in the consolidation of Grenville's leadership, but instead they marked the beginning of the most critical two weeks in his parliamentary career. The source of the trouble was again the action against the *North Briton*. Grenville may have felt justified in boasting to Lord Hertford on 28 January that 'we have got rid of Mr. Wilkes', but the issues surrounding his arrest lived on.[46] The turning-point had come on 6 December 1763 when Wilkes's case for damages came before the Court of Common Pleas. In his verdict on the arrest and imprisonment of Wilkes, chief justice Pratt declared that the general warrant used was unconstitutional, illegal and absolutely void'.[47] Though it had no immediate bearing on the expulsion policy then being planned, the judgement did strike right at the heart of the government's original legal action against Wilkes and the No. 45, throwing several of the issues arising from his arrest and punishment into doubt. In the debates of November, only Pitt had earnestly differentiated between the man and the constitutional points raised by his arrest to attack the government; the door had now been opened for a full-blown opposition campaign. If general warrants were illegal, then Wilkes should never have been arrested and the whole action against him and his press allowed to proceed.

Grenville received warning of what the opposition intended on the day of Wilkes's expulsion. Sir William Meredith (MP for Liverpool), who had left the government for the opposition over the action against the *North Briton*,

[44] Grenville had been discussing the matter with these MPs since November 1763, *Fortescue*, i. 63.

[45] *Commons Journals*, xxix. 834.

[46] *Grenville Papers*, ii. 258.

[47] *Parliamentary History*, xv. 1387.

introduced the technical motion on 'the imprisonment of the person of John Wilkes . . . and the seizing of his papers in an illegal manner' that would lead into the debates on general warrants in February.[48] There is no doubt that Grenville saw the implications of this move and how tenuous the ministry's defence of their action against Wilkes would be. In law, pending a decision in a higher court, the use of general warrants was now forbidden and no vote in the Commons could alter this fact. Not surprisingly, therefore, rather than upholding their legality, Grenville merely sought to deflect criticism on the specific case of Wilkes, and thus postpone a final parliamentary decision on their status. This line of argument just about held water. The decision in the Court of Common Pleas was not, as Pratt himself indicated, a final one. The ministry's complaint against Wilkes was still pending in the King's Bench, and until Mansfield upheld Pratt's verdict, Grenville contended that it was still the task of 'the Court of Justice and not the House of Commons to pronounce judgement on this point of law'.[49] This is exactly what Grenville told Meredith on 6 February when the latter moved for the production of the general warrant used in the arrest of Wilkes, and it received a cool reception. The judgements given by Pratt in the cases of *Wilkes v. Wood* and *Leach v. Money*, had to all intents and purposes ended the life of general warrants as a weapon of the executive. Though it named no one in person, over fifty people had been arrested on the warrant used against the *North Briton* and a growing body of MPs did not think it premature to pass a resolution confirming Pratt's opinion that this was illegal and unconstitutional. To end the debate on 6 February, Grenville was obliged to use the evasive technique of moving the previous question, and unease in the House at the government's slippery behaviour appeared in the reduction of its majority to ninety-five votes.[50]

It is unlikely that any of Grenville's colleagues would have willingly exchanged offices with him before the next stage of the opposition campaign, set for 14 February. The

[48] *Commons Journals*, xxix. 723.
[49] *Border Elliots and the Family of Minto*, p. 392.
[50] *Commons Journals*, xxix. 792.

opposition had the smell of victory in its nostrils on this issue, deluding itself into believing that here was an opportunity to overthrow a ministry by a vote in the House of Commons. The ministry could do nothing but stand its ground and wait for the bullets to fly, and for this task Grenville proved the perfect redoubt. Over the next two weeks he drew on reserves of confidence and raw nerve that drew admiration from all sides of the House. Grenville was determined to stay first lord after so carefully establishing his primacy in the successful action against the *North Briton*. The legality of general warrants represented a subsidiary issue, and he did not intend to retreat from his dominant position in the face of this opposition onslaught. As his wife wrote in her diary on 20 January, 'Mr. Grenville desired, by all means, that it should come on immediately.'[51] There was no bravado in this: Grenville assured himself from the beginning of this issue that on the procedural point in question he was in the right. A letter from John Walsh (MP for Leominister) to Grenville's friend Robert Clive in India, characterizes the aplomb with which he approached these debates:

Before the debate [on 14 February] I spoke to Mr. Grenville, and reminding him of what has passed when you introduced me to him I recalled that it was upon such occasions as the present that he had the most want of assistance from his friends, and that I was apprehensive my being no longer neutral, would, instead of being of use to him as I meant it, be of detriment, that therefore I left it to his option whether I should come down that day or not; upon which he very handsomely desired me to come down by all means and be determined by the merits of the case, and not only that day but during the whole affair.[52]

Unfortunately for Grenville, many of the MPs present at the debate on 14 February felt that the merits of the case rested with the opposition. When the government tried to end discussion of the legal status of general warrants by an adjournment motion, ministers scraped home in the division by only ten votes.[53] The scene was set for the final test of strength on 17 February.

[51] *Grenville Papers*, ii. 484.
[52] Clive MSS vol. 52.
[53] *Commons Journals*, xxix. 843.

Grenville remained unflappable on the eve of the storm. It has been said that 'the fate of George Grenville was to be decided on 17 February', but this is not strictly true.[54] Unease amongst MPs at the use of general warrants, did not, as some of the opposition believed, imply a general wish in the Commons to replace Grenville with Pitt and Newcastle. Even if he lost the vote on the legality of general warrants it is doubtful whether Grenville would have tendered his resignation. He did not see the issue as a vote of confidence, in the manner of the action against the *North Briton*. This was a special case, and when resolved, the deserters voting in the minority on 6 and 14 February were expected to return to the government. The king appreciated this point, and before the debate on 17 February made his feelings abundantly clear to Grenville: 'The Opposition might, for what he knew, carry the question of the warrants on Friday . . . but that would make no change in him in regard to his present Administration, which he meant to support to the utmost.'[55] It was one of the few moments in Grenville's time as first lord that he and the king were at one. There could not have been a better boost to Grenville's morale as he endured a torrid time in the extremely long debate on 17 February. The House was full to capacity with over 450 MPs in attendance, and preparations on both sides had been thorough.[56] For the government a formidable battery of speakers, including Grenville himself, Wedderburn, Elliot, North, and Norton, all sought to defer a resolution on the legality of general warrants until the matter was no longer *sub judice*. On the opposition side, their best speakers, Pitt, Townshend, Sackville, Conway, and Yorke, all pleaded with the House to confirm Pratt's judgement and reject this unlawful power in the hands of the executive. As the debate wore on, the oratory began to give way to interminable discussions of legal precedents and technicalities, and it was five next morning before Norton's adjournment motion came to a vote. After one of the most memorable parliamentary days of the eighteenth century, the result was a

[54] O'Gorman, *The Rise of Party*, p. 84.

[55] *Grenville Papers*, ii. 491.

[56] See, for example, Thomas Whately's letter to Grenville of 16 Feb. 1764, B.L. Add. MSS 57817A, fo. 1.

triumph for the ministry by 232 to 218 votes.[57] Grenville informed the king immediately of the outcome, concluding the letter with a line of classic understatement. 'The debate in general', he wrote after twelve hours of torment, 'was but an indifferent one.'[58]

The crisis was over. In the debates of 6, 14, and 17 February, 247 MPs voted against the government on general warrants, but of the many independents and friends of the ministry in this number, most returned to the administration fold almost at once.[59] In the calm light of day Gilbert Elliot explained to his father just what had gone on. Though the government 'were in the right in point of parliamentary and indeed constitutional proceeding', he wrote, 'such was the alarm spread of the danger of seizing papers and delegating so great a discretionary power to messengers that many friends of the government had got so far engaged in this general ground that they could not disentangle themselves, even upon our state of the question.'[60] General warrants had provided the Commons with a unique opportunity to demonstrate its independence from the court, and allow the opposition a brief glimpse of glory. The opposition leaders certainly appreciated the measure of Grenville's success in keeping them at bay. The chance to 'at least have brought the enemy to terms of capitulation', wrote the Duke of Devonshire, had been lost through 'accidents' and 'mismanagement', leaving ministers 'in possession of the field till next session'.[61] These were prophetic words, for when the opposition took up general warrants in the spring of 1765, there was no repeat of the parliamentary scenes of February 1764. On two occasions, 29 January and 4 March 1765, Grenville overcame opposition motions on general warrants by simple adjournment motions. The grounds for his action had not changed: the matter was still *sub judice* until Mansfield gave his verdict on their legality in June of that year. In the two divisions that

[57] *Commons Journals*, xxix. 846.

[58] *Grenville Papers*, ii. 266.

[59] Hardy, M.A. thesis, p. 191, n. 1. For a list of sources giving the voting patterns in these debates see *House of Commons*, i. 525.

[60] *Border Elliots and the Family of Minto*, p. 392.

[61] *Garrick Correspondence*, i. 170.

took place Grenville enjoyed majorities of 39 and 132 respectively.[62] The last act of this saga was played out the following year. In June 1765 Mansfield ended all speculation about general warrants by supporting Pratt's judgement of December 1763 declaring them illegal. This exploded Grenville's case for deferring a Commons' decision on their status, and the new ministry, led by the Marquess of Rockingham after July 1765, decided to pass the long-fought-over resolution confirming these judgements. However, Grenville still had the last laugh. When the resolution came before the Commons on 22 April 1766, the former first lord astounded everyone by moving for a bill to render all general warrants illegal. In a speech described sarcastically by one observer as 'a wonderful spring tide of liberty', Grenville argued that this was perfectly consistent with his previous position.[63] Though he had always felt personally that general warrants were illegal, the point at issue when he led the government was procedure. It had been the task of the courts to pass judgement on the legality of general warrants, and now this was complete he accepted the House must reflect these verdicts in legislation. To the Rockinghams this was not consistency but chicanery. Nevertheless, to Grenville's credit, the idea of a bill was accepted, and it proceeded to the Lords in June 1766. Unfortunately, with the dismissal of the Rockinghams the following month the bill was not taken up in the next session; but its spirit did survive. No government from that date attempted to resurrect the power of arbitrary arrest inherent in the general warrant.[64]

In the spring of 1764 Grenville stood triumphant not only at Westminster but also in the king's favour. From an issue that had promised only destruction and disarray, Grenville had emerged in an impregnable position. In his rise to the top, Grenville had displayed few of the qualities that marked out a successful leader. A faint heart and reluctance to act against his former associates had undermined his career since George III's accession. But the courage and nerve displayed first in fighting off Bute's attempts to oust him in the summer

[62] *Commons Journals*, xxx. 70 and 220.

[63] Grey Cooper to Rockingham, 26 Apr. 1766, W.W.M. R1–602.

[64] For further details on Grenville's involvement with general warrants at Westminster see Lawson, Ph.D. thesis, pp. 98–106.

of 1763, and then in the dark parliamentary days of February 1764 had eradicated all these memories. In the Commons Grenville would no longer be dependent for his dominant position on others. The trials and tribulations of these early months in office had created a hard core of followers that Grenville soon nurtured into a sizeable following. For the remaining eighteen months of his ministry, the parliamentary opposition would pose no threat at all to the government's majority. As Grenville pressed ahead with his financial and colonial legislation during 1764–5, the Commons came to view him as the man of the hour, perfectly equipped to deal with the problems arising from the Seven Years War. By the summer of 1765 Grenville was so entrenched in the office of first lord that it would take all the king's executive and prerogative power to remove him.

Chapter VI

The Months of Triumph, 1764–1765

From April 1764 to May 1765 the Grenville star remained in the ascendent. In his victory over general warrants Grenville established his primacy at Westminster, and over the next twelve months would fulfil his dream of creating his own parliamentary following. It was a remarkable transformation, and had its roots firmly in Grenville's approach to government and administration. In the words of Charles Yorke, Grenville was 'a very worthy and able man . . . whose turn lay towards the revenue, and to that public economy, which was so much wanted'.[1] In the immediate post-war years anxiety over the mushrooming national debt and high public expenditure, arising from the seven years of hostilities, became a national obsession. To modern eyes the sums anguished over seem pitifully small: it is difficult to envisage, for example, how an annual budget of £8m could cause upset. Nevertheless Grenville showed the concern of his contemporaries at the upward trend of public indebtedness, and had done so since 1759 when opposing Pitt's extravagant war policies. He did not view the unfunded debt or budgetary deficit as an integral part of the nation's economic structure, but evils to be remedied if the country was to prosper. Grenville spelt out his attitude in the king's speech which he drew up at the end of the parliamentary session in April 1763. In a clause that had the full support of the king, MPs were told that it is 'my firm resolution to form my government on a plan of strict economy. The reductions necessary for this purpose shall be completed with all possible expedition . . .'[2] Over the next two years, 'strict economy' would be the criteria by which the efficacy of all ministerial policy would be judged.

The issue that underwent the severest test by this yardstick was the post-war settlement of the American colonies.

[1] Cited in *House of Commons*, ii. 542.
[2] *Parliamentary History*, xv. 1327.

At first sight the war appeared an immense success for British hopes of expanding its sea-borne empire at the expense of the French. At the Peace of Paris, ratified in 1763, Britain obtained Canada, French Acadia, and the Floridas, taking control of all the seaboard colonies in America, and many new possessions in the West Indies. With these conquests, however, came three fundamental problems: how to assimilate these new territories into the existing imperial structure; what economic regulations to apply to the expanded empire; and who should pay for securing these newly won prizes? It is sometimes said that Grenville came to office with a programme to tackle each of these problems, so thoroughly did he set about his task of settling the colonial administration. This is an overstatement, however, for Grenville was never so predetermined in his approach. He certainly carried a knowledge of the many shortcomings in the colonial administration from his days at the admiralty and in cabinet under George III. It soon became clear too that on entering office he had the embryo of a plan on how to rectify some of these failings, especially in the area of trade regulation. But events overall did not develop in a manner that allowed Grenville to overcome problems in America by the application of a preconceived strategy. The approach he used was that well-worn pragmatic response of all eighteenth-century politicians to difficult issues; solve the problem as it arose. The principal difference between Grenville and other leading politicians of the day was that he appeared more determined to lessen the burdens of the British taxpayer once in office. He came to believe that the colonists should pay their share of providing for their own defence and administration, and enacted policies to secure this end. Grenville expressed no strong ideological commitment to this policy before he became first lord, but soon came to see the value of it after establishing himself as leader. In the first place he considered it fair to ask the colonists to make some contribution; and second, attempting to relieve the burden of taxation on Britain made him very popular at Westminster – a crucial point for a man without a parliamentary following of his own.

In the early months of office as first lord, Grenville had very little influence on American policy at all. The fateful decision to keep an expensive army in the colonies was in fact taken before Grenville became leader. On 4 March 1763 the Committee of Supply on the American and other army estimates considered and approved without dissent the plan to maintain a large military force in the colonies. This was the culmination of a complex decision-making process, which revealed mixed motives for wanting to keep the army in America. Common to those favouring retention, however, were two basic factors: a wish to keep the French from re-establishing a presence on the mainland, whether it be from discreet infiltration via the Indian nations or directly from the remaining French bases in the West Indies; and a desire to see the colonists themselves share in supporting the cost of the army. These considerations were to the fore in the committee of 4 March when the observation that 'Next Year America to pay itself' passed without comment.[3] Indeed those future champions of American liberty, William Pitt and William Beckford, complained that the force proposed was not adequate for the job at hand: 'Mr. Pitt . . . 10,000 men hardly enough to speak to one another if a communication is needed. In Canada 15 or 16,000 men as much as any the King of France's.' It is no wonder that Grenville would complain later of misrepresentation over his plans for taxing America. The House voted the army and approved the principle of a colonial contribution to its support, but left Grenville's administration to devise the means. Even if he had wanted to avoid this thorny issue of colonial taxation it is unlikely he would have been able to do so. Pressure for action was being applied as soon as the new administration took office. In April 1763, James Oswald, then leaving the treasury for a post in Ireland, presented Grenville with a lengthy memorandum on supplies for the coming year. The underlying message in this paper was that Britain's military expenses should be reduced considerably in the future. 'It seems strange,' commented Oswald, 'that a loan for so large a sum as £10,000,000 should be suggested as *necessary* after a peace concluded. When Lord Bute will recall that a loan to this amount was deemed an exaggeration even in the case of the restricted

[3] The account and quotes from the debate that follow are taken from Newdegate MSS B2543/11. For more detailed information on the decision to keep the army in America see Shy, *Toward Lexington*, chap. II.

plan of war and when it was then allowed to be necessary to keep up the whole force . . . a much less sum was insisted to be sufficient.'[4] The search for a suitable means of securing this reduction by a colonial contribution to the support of the army began eight weeks later, and would emerge in March 1765 as the Stamp Act.

Another problem over which Grenville initially exerted little influence was the attempt to devise a plan for bringing the newly acquired territories under British rule. The first policy initiative on this issue came from Egremont at the southern department, who instructed the board of trade on 5 May 1763 to put forward proposals on three points at issue: the type of governments to be established; the military force required to defend the new lands; and the best way to allow the colonists to contribute to their own defence.[5] Egremont included his own suggestions for a solution to these matters but left it to the board to draw up an overall plan. For the time being the board ignored the needs of the military establishment and how to raise revenue, concentrating in the early summer on how to bring the new territories under British rule. The cabinet discussed the board's report on this issue on 8 July 1763, at which Grenville was present. But his views on the suggestion of bringing the whole of Canada into the 'old north-west' and setting up a government there modelled on those of Georgia and Nova Scotia are unknown.[6] Not until September 1763, after the ministry had undergone a change of personnel, did the topic reappear in official business. In the interim an uprising of Indian nations led by Pontiac had prompted the board of trade to recommend immediate action on settling the boundary of the new territories. The form of action suggested was a proclamation forbidding colonial settlement in the area reserved to the Indians in order to reassure and guarantee their future security. The cabinet discussed the idea of a proclamation on 16 September 1763, and on this point Grenville had clear views, being the lone dissenting voice:

[4] Oswald MSS chest iv, B.

[5] Shortt and Doughty, *Canadian Documents*, pp. 127–30.

[6] These exchanges are discussed in full by Humphreys, 'Lord Shelburne and the Proclamation of 1763', *E.H.R.* 49 (1934) 241–64.

Some objections were made to this plan of not including all the territory ceded by the late Treaty to Great Britain in some of the governments and likewise to the line beyond which the lands *for the present* are reserved to the Indians which is to be declared by Proclamation issued *here* but as these doubts were being raised by Mr. Grenville only and against Ld. Halifax's opinion as well as that of the other Lords they were no further insisted upon.[7]

Grenville did not expand on the reasons for this disagreement in his private correspondence or diary. Nevertheless his dissent demonstrates that he had some understanding of a crucial problem facing the colonies in the post-war years. Controlling westward expansion had proved an impossible task during the eighteenth century; land speculation and friction with the Indians through trade and territorial violations by the colonists had bedevilled all attempts to create an atmosphere of peaceful co-existence. Grenville was quite right in questioning whether or not the proclamation suggested by the board of trade would eradicate these abuses. Leaving the territory occupied by the Indians outside British jurisdiction would create an administrative vacuum in which the crooked trader and avaricious speculator could still thrive: a prognosis that became an unhappy reality over the next fifteen years. It was also a mistake not to designate permanent boundaries to the native territories. This gave the impression to settlers and officials alike that the government in London was not sure of what it was doing; an attitude which in turn led to lax enforcement of the law, and encouraged encroachments on Indian lands by the colonists.[8] This is not to imply that Grenville opposed westward expansion. He simply wanted the settlement of new lands to be kept strictly under governmental control. In this manner least trouble with the Indians would occur, and the expense of employing an army to separate them from the colonists avoided. Grenville's interest in and commitment to seeing the West developed can be found in his purchase of estates in Florida, and his share in the Vandalia scheme on the Ohio river, purchased in 1770. The deal was struck with Thomas Wharton, through their

[7] Tomlinson, *Additional Grenville Papers*, p. 318.

[8] For a fuller explanation of these points see Sosin, *Whitehall and the Wilderness, passim.*

mutual friend Thomas Pitt in that summer. Grenville's enthusiasm for the project in fact was responsible for the success of the transaction; but it is clear that on its completion his hopes for the future of Anglo-American relations were tinged with scepticism:

> If the plan for this settlement in North America should survive hereafter, I am not to expect to live to see any effect from it, but everything which may be a memorial of the union of our sentiments will give me pleasure, whether that union be to preserve if possible the ancient settlement made by our ancestors in this country, or for what seems to me a less visionary purpose to establish a new settlement for our posterity in North America.[9]

Whatever Grenville's objections to cabinet policy on westward expansion were in September 1763, his colleagues received them coldly. The cabinet approved the proclamation in the form suggested by the board and the first lord was overruled. This was not a frequent occurrence. There are records of nineteen cabinet meetings during Grenville's ministry and this is the only occasion that the first lord lost a policy decision. It does not seem that he was outmanoeuvred or that anything underhand took place. The other ministers present: Bedford, Halifax, Gower, Sandwich, and Egmont, simply accepted the expert advice offered by the board of trade and southern department rather than be swayed by Grenville's more personalized vision of how policy should develop.[10] Indeed it was Halifax and his officials at the southern department who took charge of preparing and implementing the clauses of the proclamation issued on 7 October.[11] In the event Grenville probably wished he had paid closer attention to the establishment of governments in the newly acquired territories. It soon became clear that both the board and Halifax had overlooked some important legal and constitutional considerations concerning Quebec. There seems to have been a perfect ignorance of the fact that Quebec already possessed its own legal and religious infrastructure, and the government's intention to simply substitute a British model for this was neither justified nor realistic. Mansfield, who had become a confidant of the king since

[9] B.L. Add. MSS 57816, fo. 145.
[10] Tomlinson, *Additional Grenville Papers*, pp. 317–18.
[11] Shortt and Doughty, *Canadian Documents*, pp. 156–9.

Bute's departure, heard of this plan and the protests from Quebec at Court and told Grenville in no uncertain terms to re-examine the stipulations of the proclamation governing the French majority in Quebec:

> Is it possible that we have abolished their laws, and customs, and forms of judicature all at once? – a thing never to be attempted or wished. The history of the world don't furnish an instance of so rash and unjust an act by any conqueror whatsoever: much less by the Crown of England, which has always left to the conquered their own laws and usages, with a change only so far as the sovereignty was concerned . . . For God's sake learn the truth of the case, and think of a speedy remedy[12]

Grenville's response to these strictures is unknown, but it is fair to assume that he immediately referred the matter to the government's legal officers. Mansfield was revered by Grenville as the leading jurist of the time and would not have ignored his advice. Yet there would be no speedy remedy for the problem of Quebec. Ten years elapsed before a reasonable policy for governing the territory came before parliament, and the whole issue offers a clear demonstration of the shortcomings in the imperial administration before the American war for independence.[13]

In conjunction with the board's efforts at assimilating the new territories into the empire, a thorough examination of the economic circumstances in the thirteen old colonies took place. The review touched on both trade and revenue matters, and in this area of policy Grenville's influence is clearly evident. While the successes in the Seven Years War had guaranteed huge territorial gains, it had also shown 'the old colonial system' governing trade patterns between America and the mother country to be in a state of collapse. The Navigation Laws, which since 1660 set the guidelines for the transport of goods in British or colonial ships, restrictions and bounties on colonial manufactures and protected the British market for American goods, were simply not functioning as originally intended. Evidence gleaned over a number of years, but especially during the war, showed these laws were being evaded. It was not only a matter of smuggling or

[12] *Grenville Papers*, ii. 476–7.

[13] Marshall, 'The Incorporation of Quebec in the British Empire 1763–1774' pp. 43–62, in *Of Mother Country and Plantations.*

collusion with customs officers: even if prosecutions were brought against offenders, it proved extremely difficult to obtain a conviction before sympathetic colonial juries under the direction of American judges. The enforcement of the Molasses Act provided the most flagrant example of evasion. In 1733 a duty of 6*d.* a gallon on molasses imported to America was levied, but by 1762 this duty had yielded no more than £22,000. Such figures provided irrefutable proof that the system of trade regulation and revenue collection had failed miserably.[14] The illegal importation of molasses from the French West Indies had long since replaced the pattern imposed by Britain, and, more important, it was seen by the colonists as a legitimate practice. Grenville was aware of these deficiencies long before he became first lord, through correspondence with friends in America. One official in particular, Nathaniel Cotton, exerted considerable influence over Grenville's views on how to solve these problems after the war. A long letter from Cotton sent to the admiralty on 3 December 1759 demonstrates why this was so. After congratulating Grenville on the taking of Quebec, Cotton laid bare the prevalent American attitude to British trade regulations during the hostilities:

> I can only wish that the same spirit of emulation to establish the true interest of this nation appeared throughout his majesty's plantations as is very apparent here; but to the shame of some of our colonies, an illicit trade to the French islands is too much connived at in sending many supplies as tends much to keep up the spirit of our enemy, when if kept back, they must be greatly distressed.
>
> The colony of Rhode Island and some of the more southern ones are guilty of this practice to such a degree that it is almost impossible for it to be unnoticed; and as there is no officer to watch over them in their unjust proceedings, by reason of the necessary officers to inspect the entries and clearings being of their own appointment, I much fear they are such men as are too greatly concerned in the trade to prevent it.[15]

An unknown contemporary said of Grenville that he 'lost America because he read the American dispatches, which his predecessors had never done'.[16] There is a grain of truth in

[14] Barrow, *Trade and Empire*, pp. 134–57.
[15] B.L. Add. MSS 57820, fo. 193.
[16] *Albemarle*, i. 249.

this statement, for reports like that sent by Cotton were no rarity at either the board of trade or customs board. It is simply that in Grenville, officials like Cotton found a conscientious minister whose experience in law, administration, and finance equipped him to tackle these obvious deficiencies in the imperial structure. In fact it is probable that Grenville had begun to take action on smuggling and official collusion while first lord of the admiralty.[17] An Act passed in 1763 before he became first lord of treasury, and which bore his hallmark, sought to apply existing customs laws more rigorously. Customs officers were encouraged to enforce the law by the promise of at least half value of the goods lawfully confiscated. Naval ships were also empowered to seize suspected smugglers on the same terms. These amendments were circulated to the governors of each colony in early July 1763, and later that month a further stipulation, requiring all customs officials to take up their posts on pain of dismissal, was added. The thinking behind these measures betrayed no grand strategy. Grenville was shocked at the poor state of the colonial administration, and the measures could tighten discipline and increase efficiency in the customs service without any massive increase in public spending.

The next problem to be solved concerned punishment of offenders. Grenville well knew that without just retribution being seen to be brought against offenders, there would be no hope of enforcing any trade regulation at all. Grenville pursued this objective by following the recommendations of a customs board report sent to the treasury in September 1763. This document contained numerous suggestions designed to deter the illegal trader and bring him to justice, including heavier fines for smugglers, bonding of all ships in and out of American ports, and severe penalties for corruption in the customs service. The most controversial clause accepted by the treasury, however, proposed a reinforcement of the powers of the vice-admiralty courts that shared jurisdiction with common law courts in cases relating to trade and revenue. The use of these courts was an emotive issue in both Britain and America, for they sat without

[17]The account that follows is indebted to Thomas, *British Politics and the Stamp Act Crisis*, pp. 44–50.

juries and sentences handed out by presiding judges could be harsh. Yet a detailed examination of the operation of these courts in America reveals that grievances were often more apparent than real. Judges appointed to the vice-admiralty courts from the American legal profession proved just as amenable to local pressures for acquitting smugglers as the juries in common law courts.[18] To overcome this Grenville supported the proposal put forward in late 1763, and later embodied in the American Duties Act of 1764, that recommended a new super-intending court be set up under an English judge. This court would have jurisdiction over all the colonies in North America, and the site chosen was the garrison town of Halifax, Nova Scotia. If ever Grenville's narrow legal view of policy could be said to have triumphed over his common sense it was on this issue; for the remoteness of the location soon made a mockery of the decision to support the recommendation of the customs board. Even though a conviction might be more likely, the practical problems of travelling to Halifax with witnesses discouraged customs officers from using the court. While Grenville remained first lord, the acquittals of smugglers before colonial juries and judges continued unabated.

The area of colonial policy where Grenville left an indelible mark lay in the search for a revenue to support the army in America. In the spring of 1763 the board of trade had shelved this problem while it settled the governments of the new territories. But it could not be ignored for long as the decision to keep the army in America and extract some revenue from the colonists for its support had been taken before Grenville became first lord. The treasury did in fact take the initiative when in May 1763 Charles Jenkinson, secretary to the treasury, asked the customs board for an explanation of the low revenue from American customs and ideas on how to remedy this deficiency. Such a request implied the application of a new principle in trade regulation. The navigation laws were no longer to be used simply for suppressing smuggling and guaranteeing Britain's trade monopoly with the colonies but also for raising revenue. When the customs

[18] For a study of the operation of these courts see Ubbelohde, *Vice-admiralty Courts and the American Revolution*, pp. 3–71.

board replied in July they accepted this new principle and suggested the duty on molasses might be a good case for testing this new policy.[19] The existing duty of 6*d*. a gallon was artificially high and undoubtedly encouraged smuggling with the French sugar islands of the West Indies, who with the loss of Canada, were dependent on British North America for provisions and timber. A reduction of this duty would remove the need to evade the customs house and produce a revenue by the increase in the volume of legal trade. It seems that Grenville's own conversion to this idea of raising money had occurred since he became first lord, and may well have been the work of his friends and officials like Jenkinson. Charles Townshend, when president of the board of trade in March 1763, had originally put forward the idea of tampering with the molasses duty as a potential source of revenue to support the army in America. The Commons had adopted the proposal and later that month began preparing legislation for a reduction of the duty to 2*d*. a gallon. On 30 March, however, the bill was killed by a procedural device after Townshend had quarrelled with Grenville in the Commons.[20] The latter opposed outright any reduction of the duty and, presumably at this point, the new principle that trade regulations should yield a revenue. What happened between March and May 1763 to alter Grenville's mind? To judge from the correspondence at the time, he was just overwhelmed by the statistics presented by the treasury board after becoming first lord. All the financial calculations pointed to a desperate need for new sources of revenue in a time of high indebtedness, and altering the molasses duty represented one of the least painful means of achieving this end.[21]

Changing the traditional trading relationship between Britain and America was not something Grenville sanctioned lightly. Nevertheless once he accepted the necessity of raising revenue from the molasses duty he strove to fix the rate that would produce most money. Grenville resisted the reduction

[19] Thomas, *British Politics and the Stamp Act Crisis*, pp. 47–50.

[20] *Commons Journals*, xxix. 623. James Harris wrote an account of this debate and it is cited in full in Namier and Brooke, *Charles Townshend*, p. 92.

[21] Examples of these papers can be found in the Buccleuch MSS GD224/37/B.20, 6–8.

to 2*d*. suggested by Townshend, on the mistaken grounds that it would not generate sufficient income. The figure he supported, and which became law the following year, was 3*d*.; a compromise in Grenville's view between the demands of the Americans who lobbied for a 2*d*. or even 1*d*. duty and the claims of British West Indian planters who favoured a 4*d*. reduction. In the event Grenville fell between two stools: the duty was still too high to increase legal trade and create a significant revenue, and yet not high enough to satisfy the West Indian planters who saw themselves being outmanoeuvred through the continuing illegal trade with the French.[22] There was one policy, however, on which Grenville refused to give way, despite considerable pressure from within the government. It concerned the proposal to provide salaries for colonial officials from Britain. Charles Townshend had been canvassing this idea for some years before Grenville came to power, and by 1763 claimed some influential converts to the proposal. One of these was Halifax, who raised the suggestion with Grenville in January 1764. He received a very frosty reply: 'Lord Halifax dined with Mr. Grenville. After dinner they talked upon American matters, and upon the appointment and salaries of the officers appointed for the Colonies. Mr. Grenville would not consent to their having salaries from England. Lord Halifax was against this regulation, and was extremely heated and eager.'[23]

It was not difficult to see why Grenville refused to become involved in such a controversial proposal. Retaining some control over the colonial administration through the power of the purse had proved a most sensitive issue with the assemblies since the turn of the century. To attack this power in the midst of other important changes, and without warning, would be bound to cause uproar. Moreover, it would entail extra public expenditure which would more than likely be borne by the British taxpayer. Subsequent events proved Grenville's judgement sound. In 1767 Townshend became chancellor of the exchequer and implemented the legislation to secure an independent colonial administration.

[22] Johnson, 'The Passage of the Sugar Act', *W.M.Q.* 16 (1959), 507–14.
[23] *Grenville Papers*, ii. 481.

Not only did his measures cause widespread trade boycotts and dislocation but the bulk of the expense did indeed fall on the British taxpayer.[24]

The search for new sources of revenue was completed in the summer of 1763. In addition to the revenue from molasses, it was hoped to raise extra funds from duties on textiles of European and South Asian origin, and benefit from a revision of duties on wines intended to levy heavily on those imported from Madeira and the Azores. The cabinet then took the most significant step of these months when it accepted a recommendation from the treasury board for an American Stamp Act. Stamp duties were familiar to the British, having been introduced in 1671 during the reign of Charles II. The idea of applying these duties to colonial papers and documents had been raised in the 1720s, but the first serious survey on their practical application came to light in 1751. In this year Henry McCulloh, a London merchant with speculative interests in North Carolina, addressed an essay on the subject to Lord Halifax at the board of trade. From that moment on McCulloh continued a correspondence with various government departments on schemes for colonial taxation until, in July 1763, he received a favourable response from Charles Jenkinson on the proposed American Stamp Act.[25] Jenkinson asked McCulloh to draw up a detailed plan of the act, and later presented it to Grenville and the treasury board as a *fait accompli*. On 8 September 1763 the cabinet formally approved the idea. As with the reduction of the molasses duty, Grenville played no part in the initiation of the policy; succumbing to the opinions of his officials on treasury policy. Jenkinson would say later that Grenville took no active part in the preparation of the act at all:

> He said, 'that the measure of the Stamp Act was not Mr. Grenville's: if the Act was a good one, the merit of it was not due to Mr. Grenville: if it was a bad one, the errors on the ill policy of it did not belong to him.'[26]

[24] The calculations to support this case can be found in Thomas, 'Charles Townshend and American Taxation in 1767', *E.H.R.* 83 (1968), 33–51.

[25] Gipson, *British Empire*, x. 253–8.

[26] *Grenville Papers*, ii. 373n.

This declaration was not entirely accurate, however, for once Grenville accepted the need for a colonial stamp act he did keep a close eye on those drafting the bill. He was determined to ensure first, that the rates were not punitive, and second, that the revenue raised be spent in America for the support of the army. Jenkinson's statement should not be interpreted as any indication of doubt in Grenville's mind. He believed the Stamp Act to be the best and most equitable means for extracting a contribution from the colonists in support of their own defence. As he told the Commons in March 1764, stamp duties were 'the least exceptionable' means of taxation 'because it requires few officers and even collects itself'.[27] Furthermore, he found the only practical alternative, the requisition system, unacceptable in every way. This system had been in operation during the Seven Years War, founded on the constitutional principle that the king's government could 'require' each colony in time of war to contribute to their own defence. The policy proved disastrous in practice. Only those colonies directly threatened by the French complied with the government's requests for help; most contributed little or nothing at all, despite substantial reimbursement of expenses by parliament. Grenville saw the failure of requisitions in his days at the admiralty, and never forgot the lessons it taught as he piloted the Stamp Act through parliament.

The culmination of all this preparatory work undertaken by the treasury was Grenville's budget speech of 9 March 1764. 'Mr. Grenville opened the budget, fully, for brevity was not his failing', wrote Walpole, 'but he did it with art and ability too.'[28] There can be no doubting the day's success for Grenville, for James Harris echoed Walpole's sentiments when he noted in his diary that the speech 'was perfectly well heard the whole time, and gained the applause of the *whole* House'.[29] Accounts of the day's proceedings demonstrate why his resolutions were so well received by the House. In his introduction of the budget proposals, Grenville told his audience of MPs anxious about higher taxation and public indebtedness exactly what they wanted

[27] *Ryder Diary*, p. 235.
[28] *Memoirs of George III*, i. 309.
[29] Cited in Namier and Brooke, *House of Commons*, ii. 542.

to hear.[30] Government spending on pensions and the civil list had been severely curtailed; customs surveillance and collection heightened considerably and, fulfilling the declaration of intent made in March 1763, Grenville laid before the House proposals 'to raise the revenue in America for defending itself'. The centrepiece of the strategy remained the reduction of the molasses duty to 3*d*. a gallon and the proposal for an American Stamp Act; but there were many more resolutions affecting trade between Britain and the colonies. Some concerned imposition of higher taxes on imports to America, like foreign molasses; others provided bounties for colonial producers in goods like hemp and flax. Even changes in customs procedure, like the bonding of all ships in and out of American ports, that required parliamentary legislation were covered in his speech. The volume and complexity of many of these changes did in fact reflect some naïvety on Grenville's part. As Professor Thomas stresses in his detailed account of these policies, Grenville attempted to achieve too much in too short a period of time.[31] It was sufficient that he had inaugurated a change in the relationship between Britain and her colonies with the quest for revenue. A step-by-step approach from that point might have raised less protest in the colonies. Instead the shock of a new tax was accompanied by the thorough revision of the whole imperial trade pattern proposed in the budget of March 1764. Grenville did not feel he was moving too fast, however, 'This hour is a very serious one', he told MPs, 'France is in great distress at present, greater even than ours. Happy circumstance for us, as we are little able to afford another war, we have now peace; let us make the best use of it.' The intention was to revitalize a ramshackle imperial system operating to the detriment of the mother country. In its basic mercantilist form Grenville believed that America should return to its primary function of producing staple goods for British consumption in exchange for manufactured articles from the mother country. In 1765 Grenville's friend and colleague at the treasury, Thomas Whately, published a pamphlet on these measures, and it reflects the basic ideas contained

[30] The account and quotations that follow are taken from *Ryder Diary*, p. 235.

[31] *British Politics and the Stamp Act Crisis*, p. 113.

in Grenville's speech in his quest for improvement.[32] Efficient and responsible government to keep the French at bay and ensure the loyalty of the colonists to Britain, but, above all else, 'The great object to reconcile the regulation of commerce with an increase in revenue.'

The real surprise arising from the budget speech of 9 March was Grenville's decision to postpone the introduction of the American Stamp Act for a year. It has been maintained by one authority on the Stamp Act crisis that Grenville was prompted to make this decision in order to give the colonies a chance to suggest an alternative means of taxing themselves.[33] Recent scholarship has revised this interpretation of the events considerably.[34] Grenville certainly did not say in the debate that he would allow the colonies to tax themselves: he was merely introducing resolutions to the Committee of Ways and Means and not moving legislation where such a concession could be made. He changed his mind about introducing the Stamp Act after what had been said in debate. The crucial point was Grenville's statement that he would 'wish to follow to a certain degree the inclination of the people in North America, if they will agree to the end'.[35] Grenville understood the significance of introducing this new tax into the colonies and allowed for some dialogue between the government and colonial representation to take place in the final drafting of the act. This did not represent a new departure for Grenville; he followed the same procedure with respect to other colonial legislation passed during his period as first lord. He believed in the give and take inherent in any imperial relationship, and the colonial agents would have considerable success before Grenville's dismissal in July 1765 extracting concessions on both trade regulations and currency reforms passed after the March budget.[36] The stumbling block on 9 March 1764 were the objections of MPs to this procedure on such a novel issue. John Huske, a merchant of some twenty-four years' standing

[32] *The Regulations Lately Made Concerning the Colonies and the Taxes imposed upon Them, Considered.*

[33] Morgan, 'The Postponement of the Stamp Act', *W.M.Q.* 7 (1950), 353–92.

[34] Christie and Labaree, *Empire or Independence*, pp. 38–9 and Thomas, *British Politics and the Stamp Act Crisis*, pp. 72–4.

[35] *Ryder Diary*, p. 235.

[36] Thomas, *British Politics and the Stamp Act Crisis*, pp. 61–8.

in America, told Grenville that he would not be following the inclination of the colonies in any way whatsoever if he introduced the measure that year:

> Notice ought to be sent to North America of any important business which relates to them. It is done in case of Irish causes. North American agents are always desired to play for time if anything occurs in Parliament which materially affects their interest. Massachusetts Bay and another province attempted to establish a Stamp Duty but were obliged to repeal it. Would have this law read two times, printed and then sent to America for their opinion about it.[37]

This struck right at the heart of the question, for Grenville had not held extensive consultations with the colonists or their agents in London, and those in America who knew of his plans did warn that there would be trouble. As Jared Ingersoll, later agent for and then governor of Connecticut, wrote in the summer of 1763, the stamp duties would be evaded and resented because the colonists had had no say in the tax, which they believed to be their right.[38] Furthermore, it is unlikely that the government could have sent a detailed summary of the act to America by the spring or summer of 1764. Ministers had yet to consider the detailed practical problems of administering the stamp duties. An examination of Grenville's correspondence with Thomas Whately on the subject illustrates the government's lack of specialized knowledge on applying the duties in America. For 1764 their exchanges are concerned primarily with generalities, establishing areas of jurisdiction and distribution. Not until 11 April 1765 did Whately send Grenville a detailed list of all the officials appointed, which posts they would hold, and under what regulations they would operate.[39] There was thus nothing underhand or misleading about Grenville's decision to postpone the act. His mind was changed by a combination of perceptive objections in debate and simple unpreparedness on the government's part.

There is less controversy over what happend during the year's delay. The budget resolutions became law as the Plantation Act of 1764.[40] The following year Grenville

[37] *Ryder Diary*, p. 237: See also Namier and Brooke, *House of Commons*, ii. 660.

[38] *Ingersoll Papers*, pp. 294–301.

[39] B.L. Add. MSS 57817A, fo. 22.

[40] *Commons Journals*, xxix. 1027.

passed further legislation amending this legislation and adding further clauses, but they did not change the central purpose of the Act, to regulate commerce and raise revenue. In the spring of 1765 the government also rectified another anomaly arising from the decision to keep the army in America by passing the American Mutiny Act which set colonial forces on the same peacetime footing as the army in Britain.[41] The last problem Grenville sought to solve before the end of the spring session in 1764 was planter indebtedness, especially in Virginia. The solution put forward in the Currency Act of 1764 was an attempted compromise between three pressure groups: the planters who wished to retain control over the issue of their own paper currency; the Glasgow tobacco merchants who wanted stricter control on emissions of bills to safeguard the value of their debts tied up in local book credit; and the London merchants who sought to replace payments of loans in local currency by sterling. There was a mistaken belief amongst all three groups that increases in the supply of paper money necessarily lessened the value of the currency, whereas the problem was a local one in Virginia, for example, caused by the overspending on imported goods by the planter gentry.[42] But the government sought to crack the whole nut with a sledgehammer. It is apparent that Grenville and his ministers possessed only a rudimentary understanding of currency problems and, in their ignorance, forbade the issue of paper money in all the colonies south of New England, even though many of those affected by the legislation had managed their currencies well. To judge from the ease with which the Currency Act passed through parliament this initiative by the treasury would have been taken whoever held the reins of power. The conventional view of the legislation is that it represented an unmitigated disaster for the southern colonies, creating chronic shortages of specie. Recent studies show, however, that the colonists proved ingenious in avoiding the regulations, using alternative methods of exchange.[43] In effect, this was more detrimental for the British

[41] Thomas, *British Politics and the Stamp Act Crisis*, chap. 7 gives a full summary of Grenville's American policies in 1765.

[42] Ernst, *Money and Politics in America*, pp. 87–8.

[43] Walton and Shepherd, *Economic Rise of Early America*, pp. 170–3.

government as they exercised no control whatsoever over this procedure.

The completion of the currency legislation was followed by the Easter recess and Grenville's effort to solve the outstanding problem of the Stamp Act. On 17 May 1764 he met the colonial agents to explain the government's decision for postponing the act. 'His intention by this delay', wrote Charles Garth present at the meeting, 'was to have the sense of the colonies themselves upon the matter, and if they could point out any system or plan as effectual and more easy to them, he was open to every proposition to be made from the colonies.'[44] It is likely Grenville regretted uttering this statement almost at once, for it gave the impression that he was not wholly committed to the Stamp Act. Worse still, some of the agents present, like Israel Mauduit, interpreted Grenville's words as authorization for the colonies to discuss means of taxing themselves.[45] For this ambiguity Grenville must shoulder the blame. If he did envisage a colonial role in the taxation process or even an alternative measure, he should have written directly to the governors and assemblies explaining this, thus avoiding any misunderstanding. There is no way of knowing whether or not Grenville was serious about alternative methods of raising revenue. Those colonies that discussed means of taxing themselves under the mistaken idea that the government would listen to their proposals, were sadly lacking in initiative. There is a suggestion that Grenville lacked sincerity on this point with his summary dismissal of the only practical alternative to be presented while the Stamp Act was under discussion.[46] It came from Benjamin Franklin, and consisted of a plan of raising revenue by interest on bills issued in a new and uniform colonial currency. Grenville knew of the plan in 1764 and saw Franklin's final draft in February 1765, but it did not deflect him from his purpose. On the Stamp Act Grenville fell victim to a rigid prejudice about the inability of the colonies to suggest any workable alternative other than the requisition

[44] Garth was the agent for South Carolina and this quote from his letter to the colony is cited in Namier, 'Charles Garth and his Connections', *E.H.R.* 54 (1939), 646–8.

[45] *Jenkinson Papers*, pp. 305–7.

[46] *Franklin Papers*, xii. 47–61.

system, which was anathema to all British politicians. Thus when on 2 February 1765 the colonial agents did propose this, Grenville dismissed their appeals against the Stamp Act as a smoke-screen for wanting further postponements.[47]

In the atmosphere at Westminster and the country at large during 1764–5 Grenville's unbending attitude was certainly apposite. There was unanimity of opinion on the question of extracting some revenue from America to support the troops. At one point, in fact, Grenville appeared as the voice of moderation when the percentage burden the colonies should bear for defence was being calculated. 'The only difference of opinion', wrote Israel Manduit to the Massachusetts assembly on 7 April 1764, 'was that Mr. Grenville said he did not expect that America should bear more than a good part of this expense; whereas other leading members not of the Ministry said it ought to bear the whole.'[48] These feelings hardened with the news from America in late 1764 and early 1765 that the assemblies were resisting the idea of imposing stamp duties. Their immediate dissatisfaction found voice in petitions to Britain. There was no direct assertion that parliament lacked the right to levy taxation without colonial consent in the petitions from Massachusetts and New York before the board of trade in December 1764, but ministers saw this as the covert message and were stung into testing the point at Westminster. Grenville had never doubted parliament's right on this principle but went so far as to consult the foremost legal opinion of the day, Mansfield, just to make sure. Mansfield removed any lingering hesitation amongst ministers about introducing the measure as a matter of priority in the winter of 1765.[49] Indeed their enthusiasm got the better of their judgement, for on 6 February Grenville presented the stamp duty resolutions to the House without referring to the existing petitions against the measure. This was an unfortunate move, which left him open later to the justified claim that he misled the House into believing the colonies acquiesced in the proposal for stamp duties. Grenville proceeded in the face of these objections, however, because he believed the protest would die down once the tax was

[47] *Fitch Papers* ii. 324–5.
[48] Cited in Gipson, *British Empire*, x. 231.
[49] *Grenville Papers*, ii. 478.

implemented. Nobody in either Britain or America liked being taxed, he told MPs on 6 February, but 'what exemptions will [the government] go to.[?] The western country desires an exemption from cider, the northern from a duty on beer. The mischief from the 4*s*. land tax. The true way to relieve all is to make all contribute their proper share.'[50]

Such assumptions proved fatal to the future of Anglo-American relations. No one in this debate challenged the right of parliament to tax the colonies, when the division came on it was for an adjournment, not to put a negative on the stamp duty resolution. The parliamentary opposition took no serious part in campaigning against the measure and this showed up starkly in the vote which the ministry won by 245 to 49.[51] Nevertheless, it would be wrong to say that MPs approved the measure without being aware of what might result. If Grenville had wished to heed warning signs, they were certainly present in the speech by Sir William Meredith. He told Grenville

> The safety of this country consists in this with respect that we cannot lay a tax upon others without taxing ourselves. This is not the case in America. We shall tax them in order to ease ourselves. We ought therefore to be extremely delicate in imposing a burden upon others which we not only not share ourselves but which is to take it far from us.
>
> If we tax America we shall supersede the necessity of their assembling.[52]

Meredith touched a raw nerve with the Americans here, but Grenville pressed on as if ignorant of the disquiet at his taxation policy. Over the next month petitions from the colonies were dismissed for challenging parliament's right to tax, and objections to the use of the hated vice-admiralty courts for enforcing the act overruled on the grounds of expediency. The Stamp Act received royal assent on 22 March 1765, and it is easy in retrospect to attack such actions as lacking judgement and vision. But this hardly does Grenville justice. For a politician who saw the 'glorious revolution' of 1688 as the embodiment of parliament's sovereign right to govern, and therefore tax all British subjects, it never occurred to him or most of his contemporaries that the Americans

[50] *Ryder Diary*, p. 255.
[51] Walpole, *Memoirs of George III*, ii. 56.
[52] *Ryder Diary*, p. 259.

would not hold the same principles. It was absurd to Grenville for colonists to claim that their assemblies shared power with parliament rather than being a delegate authority. He did not appreciate the resentment his attempts at changing the traditional imperial relationship would arouse, and yet he was in good company. When the Stamp Act became law, there was something of a scramble for posts in America to administer the new duties, and amongst those seeking favours were the colonial agents, like Benjamin Franklin, who had recently been telling Grenville not to introduce the legislation. Who could blame Grenville for not forseeing the tide of protest about to burst forth, if it escaped the notice of the agents? Protest or not, Grenville did not believe at root that any of his American legislation was asking an unreasonable price for the continuing loyalty of the colonists. The Stamp Act, in particular, would pay only one-third of the army costs in America, the money would remain in the colonies and the whole would be administered by native-born American colonists rather than appointees from Britain. There was nothing malicious in Grenville's attitude to America in the post-war period. He was motivated by an honest desire to bring order to the newly expanded empire; the failures stemmed from attempting too sudden a renovation of a structure weakened by years of salutary neglect.

It is ironic that on policies for which Grenville was widely praised at the time, historians have since found so much to dispute. Many of the problems of interpretation arise because the colonial policies are taken out of context. Grenville brought order to all aspects of government policy and administration, of which American issues were just a part. Apart from his grander efforts at reducing the unfunded debt and reform of the revenue, Grenville attended to the details of government administration which were so much appreciated by officials. In September 1764, for example, Sir James Porter, minister to Brussels, told a friend that 'Mr. Grenville will deserve a statue from all the king's servants, especially those abroad. I see we shall now be paid regularly.'[53] Not since Henry Pelham was first lord had the country enjoyed

[53] H.M.C. 12th report, ix. 342.

the benefits of a leader who was so concerned about the taxpayer's money. Here was a first lord whose public utterances about economy and efficiency could be matched by private actions. When he told the king in May 1765 of his achievements in reducing pensions and secret service money it was no empty boast. Grenville had the figures to prove it, and a whole year spent saving money on contracts, insisting officials attend their duties and hard bargaining over appointments to support that.[54] This involvement at the grass-roots level of ministerial business was the rock upon which Grenville's popularity at Westminster rested. Robert Clive's friend, John Walsh, put the common view in a nutshell when he wrote on 3 July 1764, 'in all removes Grenville must be employed for he is indefatigable in business and they cannot do without him'.[55]

One area of policy in Grenville's ministry that has suffered neglect in comparison to the attention given America, is the approach to foreign affairs. This is unfortunate for its sheds a new and interesting light on Grenville's attitude to colonial policy. Sir Lewis Namier detected a connection between the two when he stated that Grenville only concerned himself with foreign policy when it 'bore on finance'.[56] But the implication in this statement that Grenville otherwise played little or no part in foreign policy decisions requires revision. The fact was that all foreign policy decisions since Walpole's premiership had touched upon finance because the British had built alliances with European powers to curb French expansionism and defend Hanover. The bricks and mortar of these alliances were British subsidies. No matter who was first lord, or whether or not he favoured a Prussian alliance, like Pitt, or an Austrian alliance, like Newcastle, the treasury had to sanction the subsidies, and by necessity express preferences and opinions on overall policy. In this process Grenville proved no exception to the rule. Yet his rôle did not begin and end with mere consideration of the accounts. The records of cabinet meetings during Grenville's two years as first lord show he attended all the important gatherings deciding policy, and played a leading part in formulating

[54] *Grenville Papers*, iii. 143–5 and n.
[55] Clive MSS vol. 52.
[56] *House of Commons*, ii. 542.

diplomatic strategies.[57] This came as no surprise to his colleagues. Grenville did not arrive at the treasury ignorant of foreign affairs or devoid of ideas on how British interests abroad should be directed and protected. On the contrary he had established his views quite clearly in the winter of 1761 when he broke with Pitt, and nothing occurred before April 1763 to change them. His desire to end subsidies to European powers and dispense with the eternal concern over Hanover remained firmly intact; as did the drive to expand the sea-borne colonial empire, of which America formed the core. Inherent in this approach was a change of emphasis in British attitudes to the two Bourbon powers, France and Spain. No longer would the diplomatic priority be the creation of equipoises in continental Europe to maintain the status quo. The real diplomatic success would now be won at sea, in the maintenance of British naval supremacy, and through this the colonial empire. Britain had emerged from the Seven Years War with vast territorial prizes, and Grenville knew the French would not rest under Choiseul until the balance had been redressed in favour of the Bourbon powers. To avoid this Grenville's ministry developed a strategy of devastating simplicity – a mixture of hard bargaining and gunboat diplomacy. In this approach Grenville found unexpected but powerful allies in the king and public at large. Though his policies were not formulated deliberately to court popularity, his 'little England' strategy matched the mood of the country after a long and expensive war.

This attitude to diplomacy is frequently represented as the birth of a classic period of British isolationism, ending only when the French revolutionary armies marched across Europe in the 1790s. Yet there was less intent and more accident about the strategy during Grenville's period as first lord. When he came to power the government had already disentangled the country from the Prussian alliance, and, as recent diplomatic studies have shown, this was no desertion of a faithful ally. The devious antics of Frederick the Great, and the changing loyalties of Austria and France on the Continent, rendered an Anglo-Prussian treaty meaningless on its wartime basis.[58] What gave Britain's momentary

[57] Printed in Tomlinson, *Additional Grenville Papers*, pp. 317–38.

[58] See, for example, Spencer, *Sandwich, Diplomatic Correspondence*, pp. 10–20.

isolation after the war a look of deliberate policy were two decisions by George III and Grenville in 1763–4, that in reality bore no relation to each other: not to seek alliances for the defence of Hanover or involve Britain in another German war, and avoid involvement in the Polish question that simmered over the summer of 1764, threatening peace in Eastern Europe. The willingness to sacrifice Hanover in a future European power struggle was an emotive rejection by George III of the driving-force behind his grandfather's diplomatic policies. The young king, who gloried in the name of 'true Briton' and once referred to the Electorate as that 'horrid Hanover', wished to follow the maritime colonial policies favoured by Grenville. They spoke 'very confidentially upon all his business and particularly foreign affairs', noted Mrs Grenville in her diary, 'in which he [the king] said he was persuaded Mr. Grenville and he were of a mind.'[59] Another sovereign might well have taken a different view of Hanoverian interests, but to George III and Grenville there was no doubting the wisdom of this decision. They represented a school of thinking on foreign policy stretching back into the seventh century when the anti-ministerial Whigs had campaigned against William III's plans for a large standing army to assist in maintaining the balance of power in Europe. In their eyes the policies pursued under the first two Georges were detrimental to Britain's traditional role in European diplomacy, which they meant to redress.

The refusal to become involved in the election of a new king in Poland was founded on quite different motives; as Grenville explained to the king in December 1764:

> Mr. Grenville went to the king, to speak to him about drawing his Speech, and mentioning to His Majesty that it would be proper to take notice of the election of the king of the Romans, and the king of Poland: His Majesty from thence talked a good deal upon the present situation of Europe. Mr. Grenville took occasion to observe to him how advantageous it had proved His Majesty's having stood entirely neuter, without the expense of a shilling of money to bring about either of those desireable events, and begged His Majesty to remember it had even been his advice to him to let his mediation and assistance be courted by all the Powers in Europe, rather than offered; that this advice he gave him, both as to foreign affairs and those of his own

[59] *Grenville Papers*, ii. 240.

kingdom, never to seek others, but to put himself in such a situation as should make him be sought; that it had been his opinion in regard to Prussia; it was equally so in regard to all the other Powers.[60]

This is a more definitive statement of isolationism; but it is a pragmatic one, rooted more in Grenville's tight economic approach to government than strategic expertise. It was well known in government circles that Grenville not only wanted to keep public spending down but also opposed any subsidy policy to European powers. The Polish issue fell into both these categories and Grenville never dreamt of going before the Commons to seek approval for payments to parties in the Polish election. Notwithstanding such intransigence on the point of expense, Grenville did not object to other forms of involvement. A great deal of effort was spent during his leadership in trying to cement a commercial agreement and formal treaty of alliance with Russia.[61] Negotiations on the former were important for a guaranteed supply of naval stores, and talks begun under the Earl of Buckinghamshire continued in fits and starts until agreement was reached just before Grenville's dismissal in July 1765. The treaty of alliance foundered, initially on a proposal committing Britain to assist Russia in the event of Turkish aggression. Grenville believed that such a clause was both strategically unsound, as it required directing naval vessels from the defence of colonies, and potentially expensive, for subsidies at some juncture would certainly be necessary. These unspoken conditions proved insurmountable in further discussions, and the alliance treaty never materialized, in spite of the very strong desire to acquire one. Grenville also declined to veto Sandwich's attempts in 1763–4 at negotiating an alliance with Holland and Austria, along traditional lines of curbing French expansionism and defending Hanover. He took the view that the negotiations were purely a departmental matter until Sandwich presented the cabinet with any concrete proposals. The talks came to nothing in the end because the Austrians felt their interests now lay in a French alliance, and it was probably as well for Sandwich that it finished in this way. Any threat of war or

[60] *Grenville Papers*, ii. 532–3.

[61] The account that follows is based on the analysis of foreign policy in these years by Spencer, *Sandwich, Diplomatic Correspondence*, pp. 25–6.

demand for payment to support the alliance would have brought a round condemnation from a majority in the cabinet.

In fact it is the activities of Sandwich and his fellow secretary of state, Halifax, that have led to the mistaken belief that Grenville only concerned himself with foreign affairs when they touched upon finance.[62] The basis of this impression is the public quarrel in the summer of 1764 between Grenville and his leading secretaries over the government's attitude to the tardiness of the French in settling the terms of peace. There were several matters at issue, all bearing on finance; but the principal clauses disputed were outstanding debts arising from the maintenance of French prisoners; redemption of French money issued in Canada during the war; and the destruction of the fortifications at Dunkirk. The underlying cause of the argument was Grenville's insistence that the French be made to honour the time-scale for complying with the articles of peace laid down in the final treaty. He wished to deal harshly with France, ensuring that it remained weak in both Europe and the colonies; as this was in Grenville's eyes the only certain means of guaranteeing British supremacy. Halifax and Sandwich differed only by degree, in that they feared driving such a hard bargain would, out of humiliation, prompt the French to declare war. They enlisted the powerful support of the Francophile, Bedford, and with his help overcame Grenville's attempts at forcing the French into a corner over the articles of peace. They succeeded in extending the time-scale for settlement of the debts and Canada bills, while also persuading Grenville against the use of force against Dunkirk. This difference of opinion which became very heated and may even have threatened the stability of the ministry was, however, Grenville's only set-back. On his general strategy of immediate response to any threat against the colonies, the cabinet was at one. The instrument of this policy was the British navy, and the beauty of the strategy lay in the fact that a small, inexpensive fleet could secure British imperial interests after the war by a minimum show of force.[63]

[62] Ibid., p. 64.

[63] These developments can be followed in *Grenville Papers*, ii. 344–418.

The precedent for this action by Grenville's ministry was set in the early months of 1764 when the cabinet considered the problems that the East India Company faced with native and European powers in India. Grenville knew all too well how important the fortunes of the Company were to the country, for in the summer of 1763 he had rescued the Company from a credit crisis that threatened its very survival, and with it much of the financial structure in Britain.[64] He demanded therefore that the government throw its weight behind military and diplomatic efforts to re-establish British supremacy in India. The first stage was to secure the return of Grenville's friend, Robert Clive, to India as commander-in-chief in the spring of 1764. Grenville recognized in Clive a brilliant organizer and military leader; and triumphs on the battlefield were the likeliest solution for the Company that had gained much from the Peace of Paris of 1763 but had since been unable to guarantee its territorial possessions. It proved a masterstroke. Within two years Clive had turned the tables on the French and native forces, securing Bengal for the British and establishing supremacy in the Madras and Bombay presidencies. Against the Dutch, Grenville used only the threat of naval action to achieve his ends. In February 1764 the cabinet discussed a complaint from the Company that the Dutch East Indies Company was attempting 'to break up a new settlement at Salanhetty supposed to be made by the English East India Company . . . to interrupt the new passage and navigation lately discovered round the Philippine Islands to China'.[65] The directors requested naval assistance to deal with the Dutch but Grenville believed diplomacy would be sufficient. He judged correctly, for representations by Sir Joseph Yorke in the Hague brought apologies from the Dutch and the dispute was resolved amicably.

This success in the East was matched by comparable successes in Africa and the New World against the Bourbon powers. The first flashpoint occurred in the winter of 1763–4 when the Spanish expelled British log-cutters from Honduras in contravention of article 17 of the Peace of Paris.

[64] For a fuller account of Grenville's involvement with the East India Company and Clive in the years 1763–5 see Lawson, Ph.D. thesis, pp. 233–52.

[65] Tomlinson, *Additional Grenville Papers*, p. 319.

Governor Lyttelton of Jamaica informed the government in June 1764 of the action and from the beginning Grenville urged strong measures to protect the rights of the log-cutters. In this case, as it was so near to the precious West Indian and American colonies, he advised the use of a naval squadron even before negotiations were completed. Sandwich and Halifax at first thought this rash, but Grenville argued that it was in Britain's best interests, for to show any weakness in this theatre would only lead to further encroachments by the Bourbon powers. As Mrs Grenville noted in her diary on 27 September, 'The Spanish Ambassador here finds the Secretaries of State much less tenacious than Mr. Grenville, who he says, would lose all he had in the world rather than suffer diminution of the honour of the king his master, or of the commerce of the kingdom.'[66] Grenville had his way in the end because of news of French aggression in the West Indies arrived in London during the same month. British subjects and property on Turks Island had been attacked in an effort to drive the settlement away. There were grounds for the French complaints against privateering from the island, but Grenville refused to listen to such claims. He again advised swift naval action, hand in hand with negotiations, and the cabinet swung behind him. Not only did ministers fear giving any comfort to the French but they also wished to avoid criticism at home. Their fears were expressed by Sandwich in a letter to Grenville of 8 August 1764:

> I am sorry to say that I agree with you entirely in your apprehensions about the affair of Turks Islands, something of the sort you propose must be done, otherwise the clamour will be great and universal; but I doubt whether Sr W. Burnaby must not be reinforced as he has no ships of the line with him unless you so call a bad 50 gun ship with her lowest compliment of men and if you send reinforcement from hence, I fear it would operate upon our funds, but of that you are the best judge.[67]

Grenville was indeed the best judge, for in taking this tough stand and not concerning himself with the expense, he forced the French and Spanish governments to back down. Both incidents were resolved in Britain's favour and without

[66] *Grenville Papers*, ii. 516.
[67] B.L. Add. MSS 57810, fo. 115.

loss of face. This firm policy of gunboat diplomacy was completed successfully early in 1765 when the government ordered ships to Gambia in West Africa to protect British trading interests against an expanding French post at Albreda. Fortunately an incident was avoided by the diplomatic efforts of Sandwich and Grenville, the French accepting the British view that they were in Gambia on sufferance and in no circumstances should they interfere with existing trade patterns.[68]

There can be little doubt that Grenville's tough policy against the Bourbon powers was a great success in the immediate post-war years, maintaining British supremacy at sea and in the struggle for colonial empire. The French in particular were bent on a policy of *revanche* after the failures of the Seven Years War, and the targets, as Grenville suspected, would be British colonial possessions. In his use of the navy to defend Britain's imperial interests and a policy of hard bargaining Grenville provided the perfect antidote to French aggression. Moreover his approach to foreign policy with its concentration of maritime colonial measures complemented the work done on the internal reorganization of the American colonies well. Grenville saw in the colonial empire a certain base for the Britain's future glory and prosperity. The American legislation of 1764–5, in his eyes, brought order and efficiency to the imperial structure, while the gunboat diplomacy and hard bargaining guaranteed the permanence of these structures against the designs of France and Spain. Little did Grenville know as he surveyed his year of triumph in the spring of 1765, that it was soon to crumble into a bitter personal defeat.

[68] Tracy, 'The Gunboat Diplomacy of the Government of George Grenville, 1764–1765: The Honduras, Turks Island and Gambian Incidents', *H.J.* 17 (1974), 711–31.

Chapter VII

The Fall: from First Minister to Opposition Leader, 1765–1767

Grenville's fall from power was as sudden as it was unexpected. In February and March 1765, the king fell victim to a physical ailment that kept him confined to St. James's Palace, and restricted his ability to carry out official duties. Grenville did see George III during this time and there was not too much dislocation of ministerial business; but there was a notable change in the king's attitude towards his first minister. Since the turn of the year a coolness had developed between George III and Grenville, which the latter did not take too seriously at first. The king's incapacity, however, seemed to widen the rift. After one meeting on 18 March, for example, Mrs Grenville noted in her diary that her husband 'found the king's countenance and manner a good deal estranged'.[1] Relations worsened over the next month after the king asked the cabinet to formulate a regency bill in case of the sovereign's death. There was nothing awkward in this; George III's grandfather had taken such steps in 1751; but as the bill progressed through parliament a poignancy appeared in the entries to Mrs Grenville's diary. A fatalism emerged: Grenville could see he had lost the king's confidence and felt there was little he could do to retrieve the situation. The roots of their estrangement ran deep, and by June there was no hope of bridging the gap. It took the king two attempts to rid himself of the ministry whose 'whole attention was confined not to the advantage of their country but to making themselves masters of the Closet'.[2] By July he had succeeded, and Grenville spent the last five years of his life in opposition.

Much has been written on the role of the Regency Bill in the demise of Grenville's administration, and most recent

[1] *Grenville Papers*, iii. 122.
[2] *Fortescue*, i. 164.

scholarship suggests it was no more than a catalyst in the king's final decision to find a new leader.[3] In his own version of the changes that took place in the summer of 1765, the king's motives for dismissing Grenville are revealed as a mixture of personal and practical considerations. George III had had enough of Grenville's 'insolence' in the Closet, and wearied of battling over each new appointment in the government. 'No office fell vacant in any department', the king wrote from bitter experience, 'yet Mr. G. did not declare he could not serve if the man he recommended did not succeed.'[4] The incidents George III remembered certainly bore out his perennial charge that Grenville did not maintain the dignity of his office by involving himself in appointments that were traditionally none of his concern. In an extremely bitter quarrel over the replacement of the Court painter on Hogarth's death in 1764, the king suffered considerable humiliation and embarrassment after deciding to abolish this royal sinecure. Grenville sent for the surveyor of works, wrote George III, and abusing him for 'my having curtailed the painters office . . . he used this very remarkable expression, that if men presumed to speak to me on business without his leave that he would not serve an hour; had I followed my own inclinations I certainly should have dismissed him the moment I heard this.'

What could have prompted Grenville into such a rash and insensitive act? Not surprisingly he viewed the problem of patronage quite differently from the king. There was in his eyes no simple division between royal and ministerial appointments. Grenville pursued the control of all patronage from a lack of security. Despite the declarations of loyalty by George III in the summer of 1763, Grenville did not trust the king, nor did he believe that Bute had retired from his rôle as minister behind the curtain. In retrospect these seem irrational prejudices. Bute had retired when requested to do so in 1763, and in the months that followed George III gave several demonstrations of his good faith in Grenville, not least during the general warrants debates. After the

[3] See J. R. G. Tomlinson, M.A. thesis, pp. 165–90; Jarrett, 'The Regency Crisis of 1765', *E.H.R.* 85 (1970), 282–315 and Brooke, *King George III*, pp. 110–13. There are in these accounts, however, differences of opinion over detail.

[4] This and the quote that follows are taken from *Fortescue*, i. 164.

double-dealing by George III and Bute in the summer of 1763, however, Grenville never exorcised the evil spirit of secret influence from his mind. Bute returned to London from the country in March 1764 for no other reason than he was bored with life in Bedfordshire but all Grenville's fears and suspicions about his hold over the king returned. They were even reinforced by the feelings of other members of the government, especially Bedford whose dislike for Bute surpassed that of Grenville. On 8 July 1765 he wrote to Grenville that 'I wait for my letter of dismission with impatience, though I can hardly bring myself to believe that any people will be hardy enough to undertake an Administration, which is constructed on no better foundation than the support of Lord Bute's favouritism.'[5]

Thus in Grenville's version of the struggles over patronage there was substance to his belief that Bute's influence in the Closet remained paramount. He objected in particular to the many instances of the king claiming the right to appointments around his person, then ignoring Grenville's advice on the matter and promoting someone connected with Bute. Sir William Breton's appointment as keeper of the privy purse in 1763, was the first of several aggravating incidents that followed this pattern. The most contentious issue of all, however, concerned the control of Scottish patronage. Since Grenville's appointment as first lord, Bute's brother, James Stuart MacKenzie, lord privy seal of Scotland, had controlled patronage north of the border. This arrangement might have worked satisfactorily if Grenville's relations with the king had been good but it simply acted as an irritant, providing further cause for discontent. After one or two skirmishes with MacKenzie in 1764 and early 1765, when Grenville's recommendations were ignored in favour of Bute's followers, the matter came to a head in March when they quarrelled over James Duff's claim to the office of registrar of the Order of the Thistle. Duff was the brother of Lord Fife, faithful follower of Grenville at Westminster, and the post was to be his reward for loyalty. Grenville met MacKenzie on 25 March and recommended Duff for the job, in a manner that brooked no contradiction.

[5] *Grenville Papers*, iii. 70.

Mr. Grenville recommended Mr. Duff, Lord Fife's brother, to Mr. Mackenzie, to hold the office of Registrar to the Order of the Thistle. Mr. Mackenzie told him he had many applications for that office, but that he would lay Mr. Duff's pretensions before the king. Mr. Grenville answered, that he, Mr. Grenville, could do that; but what he desired of him (Mr. Mackenzie) was, to back those pretensions, and really to prefer Mr. Duff at Mr. Grenville's desire.[6]

Here in a nutshell are the irreconcilable differences of opinion over how to conduct the government's everyday business that finally lost Grenville the king's support. Yet in the argument over Scottish patronage Grenville had good cause for complaint. No other minister would have tolerated a portion of his appointments being held by someone outside his control. Mackenzie refused to be browbeaten, however, and with George III's obvious blessing appointed Bute's friend Sir Henry Erskine to the post. This episode proved sufficient to convince Grenville that Bute continued to exert a secret and unconstitutional influence over the king, and, moreover, would use it to bring his administration down.

The predictable result of these events was a deterioration and embitterment in Grenville's relations with the king. A solid basis of trust had never really existed between the two since August 1763, but in the months of hectic legislative activity that followed they evolved a means of carrying on ministerial business that avoided strife. During the king's illness of February and March 1765 this tenuous working relationship dissolved. George III found it impossible to keep his personal detestation of Grenville in check. The latter's endless sermons and discourses continued unabated during the king's incapacity, until George III could stand no more. On the Regency Bill the king's patience ran out. After Grenville discovered that he had not been the first to be asked to give his views on the bill, a stormy meeting ensued. 'He loudly complained to me of want of confidence for not having consulted him on this', noted the king, 'I treated his suspicions with contempt.'[7] Once the Regency Bill became law the king resolved to make soundings for a new ministry. In his own reflections on these events George III laid stress on his dissatisfaction with policy and patronage

[6] *Grenville Papers*, iii. 124.
[7] *Fortescue*, i. 165.

for seeking to replace Grenville, and yet this was only a smoke-screen. No minister could have done more in terms of legislation and securing a large majority at Westminster than Grenville. The simple truth was that George III could not stand the sight of Grenville in the Closet, and squabbles over patronage provided a ready made excuse to be rid of him.

Grenville awaited his fate philosophically, and during March and April concentrated on the preparation of the Regency Bill. What concerned him so soon after Duff's affair was the king's wish to keep the nomination of the regent a secret. Like many of his contemporaries, Grenville put this reticence down to a firm intention on the king's part to name the princess dowager as regent, opening an obvious connection with Bute. It is now known that George III intended no such thing. He saw the Regency Bill as a family matter, and declined to name anyone in public because his favouring the queen for regent to the exclusion of the Duke of York would have caused a bitter family row. The king desired the nomination to be kept secret and only to become known if the regency came into force – if George III lived York would never know of the exclusion.[8] He said nothing of his motives to Grenville, who seethed throughout the bill's passage under the mistaken belief that Bute had again exerted his evil influence over the king. Though George III's inexperience over the issue hardly did justice to Grenville, none of his pleas to the king for an honest answer to the question of who would be regent received an answer. In fact Grenville's sermons on the need to reveal all, merely brought forward the king's decision to seek a new leader.

It was Bedford who first realized that something was afoot. He saw the king on 9 May, taxing him with 'the reports which were got about of an intended change'.[9] He received no reassurance from George III's reply, and his suspicions were borne out five days later when the Duke of Cumberland began negotiations with Pitt about forming a new government. Their talks took place against a background of violent riots in London by distressed silk weavers, angry

[8] Brooke, *King George III*, pp. 110–11.

[9] The following account and quotations are taken from Mrs Grenville's diary, *Grenville Papers*, iii. 159–71.

at the Duke of Bedford's role in blocking a proposed rise in the duties on imported silks. They attacked Bedford's London residence and it took troops three days to restore order. It proved a turn for the worse for Grenville, who had been in favour of restricting imports. In an interview with the king on 16 May, Grenville was left in no doubt that George III held him partly responsible for the riots and the fact that they had taken so long to quell. This was too much for Grenville to bear, when he knew that negotiations were going on behind his back. At their next meeting on 19 May Grenville made an audacious attempt to draw the king into the open: 'He [George III] then signed the papers Mr. Grenville brought to him, and was going to have bowed him out of the room when Mr. Grenville told His Majesty that he believed he had forgot to give him orders relating to the change of his Government.' Grenville's 'insolence' was so commonplace in the Closet that the king probably never gave this statement a second thought. He was so confident of the negotiation with Pitt succeeding that he more or less admitted that changes would be taking place very soon. Grenville left the audience believing that he had delivered his parting shot as first minister.

The king should have known better than to rely on Pitt. Despite a personal visit by Cumberland to his home at Hayes, the great man refused the king's invitation, and Newcastle's friends in opposition declined to serve without him. The stumbling-block to Pitt's acceptance appeared to be Temple's refusal to join the projected administration. Pitt needed his brother-in-law as right-hand man because of the poor state of his health.[10] Temple's reasons for rejecting the call to office were never stated openly, but concerned matters of a private and sensitive nature. In practice these turned out to be an unexpected reconciliation with his younger brother, George Grenville, sealed on 21 May at a well-publicized meeting. Those responsible for this bombshell were not only mutual friends, like Augustus Hervey (MP for Saltash) but also other members of the family. James Grenville in particular grieved over the family divisions caused by political disagreements, and even though

[10] H.M.C. Stopford-Sackville, i. p. 21.

he adhered to Pitt, he spent the spring of 1765 supporting efforts to reconcile the various factions.[11] Once Pitt realized James Grenville's commitment to achieving this end, he exerted no pressure on Temple to join the planned administration. At root neither wished to become involved in a campaign to undermine Grenville's position simply to satisfy the king's personal vendetta. As Lord Frederick Cavendish, observing the situation, put it, Pitt 'drew a conclusion from the situation of the present ministers, that if they were turned out for no other reason than supporting the measures they advised, it *augured* ill for him, and therefore he must know why they were turned out'.[12]

This left the king with the worst possible alternative, 'the most cruel of all necessities, the keeping those men that I thought neither from the weight, abilities, nor dutiful deportment worthy of their stations'.[13] George III summoned Grenville on 21 May and asked him to continue, expecting an immediate acceptance. He did not receive it. Grenville replied that he could make no decision until he consulted other members of the cabinet, which met later that day. There is no doubt that Grenville saw an opportunity during this impasse to humble the king, and became carried away with his own self-importance. In a letter to his friend Lord Strange on 23 May he likened the king's dilemma to the situation in August 1763 from which his ministry emerged so victorious, and all the portents seemed to indicate a repetition of this success.[14] As Whately reported later that day, feeling at Westminster ran strongly in Grenville's favour; 'I have heard the sentiments of many people to-day and they are all entirely agreeable to what your best friends must wish them to be: I find some determined to support you whom you did not count upon.'[15] Thus when the cabinet laid down its terms for continuing, the king's feelings and opinions received rough treatment. They consisted three principal conditions:

[11] See, for example, B.L. Add. MSS 57807, fo. 26.
[12] W.W.M. R1–449 (21 May 1765).
[13] *Fortescue*, i. 165–6.
[14] Grenville Letterbook, vol. 2.
[15] B.L. Add. MSS 57817A, fo. 24.

1st. That the king's ministers should be authorized to declare that Lord Bute is to have nothing to do in His Majesty's Councils or Government, in any manner or shape whatever.

2nd. That Mr. Stewart Mackenzie be removed from his office of Lord Privy Seal of Scotland, and from the authority and influence which has been given to him in that kingdom.

3rd. That Lord Holland be removed from the office of Paymaster General, and that office disposed of as has been usual in the House of Commons.[16]

After hearing these distasteful conditions on 22 May, the king himself asked for time to consider: as he wrote afterwards, 'I could not be so wanting to myself as to treat them otherwise than as jailers.'[17] He particularly resented the clause on Mackenzie's dismissal, for the post as privy seal in Scotland had been promised for life. This response was not what Grenville had expected. After reporting to his colleagues, Bedford, in fact, told him that they had gone too far. As the king intended to 'give his answer to you singly', he wrote, 'I conceive . . . that everything is over'.[18] Grenville agreed, regretting that the king still wished changes 'which I fear can be of no advantage'.[19] They were somewhat taken aback therefore when, later that day, the king agreed to eat humble pie and allow them to continue on their terms.

It proved a phyrrhic victory. A subtle difference existed between the king's capitulation in August 1763 and that of May 1765 – there was still a means of escape. Grenville should have realized that Cumberland's involvement meant Bute had retired from politics altogether. The king's uncle hated Bute as much as Grenville or Bedford. Instead they made the mistake of viewing Cumberland as yet another evil adviser to be separated from George III. The Newcastle Whigs appreciated the significance of Cumberland's involvement in the negotiations, however, and it was they who reaped the benefit. Of Grenville and Bedford's behaviour after 22 May, nothing complimentary can be said. In attempting to ensure that the king never went behind their backs again, they merely added insult to injury. Before retiring for the summer recess to their country homes both

[16] *Grenville Papers*, iii. 41.
[17] *Fortescue*, i. 166.
[18] *Grenville Papers*, iii. 44.
[19] B.L. Add. MSS 57811, fo. 27.

delivered sermons to the king, reminding him of 'the little countenance he showed to his Ministry, and how difficult it was for them to go on under such difficulties'.[20] Wiser men would have left matters alone after 22 May. George III's biographer, John Brooke, describes the deep and lasting impression the events of these weeks made on the king's subsequent political behaviour.[21] Never again would he suffer such humiliation, allowing his ministers to dictate the terms of service, and he would do all in his power to keep Bedford and Grenville from taking office again. In this atmosphere the king needed little prompting from Cumberland to reopen negotiations for a new administration. On 17 June, the day Grenville left London, the Duke of Grafton visited Pitt to talk about forming a new government. They quickly reached agreement on places and policies, for, like Newcastle, Pitt knew Cumberland's involvement meant Bute's retirement was final. The discussions never reached fruition, however, because Temple again refused to come in. Not only did he dislike the role of figure-head assigned him in the projected administration, 'the plan of the provisional administration,' he told Grenville, 'was, I think, Butal-Ducal', but he also shrank from the idea of becoming an instrument of revenge on his brother now they were reconciled.[22] The king had prepared for this rejection, authorizing Cumberland in case of failure to deal direct with Newcastle and his friends. Through persistence and two weeks of hard bargaining the king succeeded in persuading them to take office without Pitt. On 10 July 1765 Grenville handed in his seals of office.

Grenville's immediate reaction to his dismissal was passive anger. The account of his last audience of the king, where they ran over the ground of past disputes, is a catalogue of a minister sensing betrayal at every point, and their relationship finished as it begun in 1763.[23] Grenville knew in the first week of July that the new ministry, under the leadership of the Marquess of Rockingham, intended to reverse many of his policies, and he lectured the king on the catastrophes ahead if he allowed this to happen:

[20] *Grenville Papers*, iii. 194.
[21] *King George III*, pp. 118–22.
[22] *Grenville Papers*, iii. 64.
[23] The following account and quotations are taken from ibid., pp. 215–19.

Mr. Grenville told him he understood that the plan of his new Administration was a total subversion of every act of the former; that nothing having been undertaken as a measure without His Majesty's approval, he knew not how he would let himself be persuaded to see it in so different a light, and most particularly on the regulations concerning the Colonies . . . that if any man ventured to defeat the regulations laid down for the Colonies, by a slackness in the execution, he should look upon him as a criminal and the betrayer of his country.

This was quite a parting shot, and it was meant to be just that. It seems that Grenville saw this bitter end to an unhappy relationship as the prelude to a lengthy retirement from the political scene. He did nothing during the summer to indicate that he was interested in organizing or partaking in a campaign against the new ministry. The last thing on his mind was putting himself forward as an opposition leader at Westminster. 'Every man was the best judge of his own situation', he told Charles Yorke on 18 July.

The reaction of his friends and the behaviour of the Rockinghams in office soon changed all this. During his term of office Grenville had acquired a personal following whose loyalty went beyond the perquisites of office. Many of these had no intention of retiring to the country for a quiet life, and, more important, they were insistent that Grenville should not do so either. These MPs not only desired Grenville's leadership they expected it in view of the changes that had taken place. Sandwich summed up the situation to his former leader on 15 August:

I have seen lately different people and of different parties and ideas in many things, but all agree in one language and that by no means a favourable one with respect to the present administration, who I think will never have the weight, which you seem to hint at, thrown into their scale and consequently their situation must be very precarious indeed if it depends upon that event.[24]

This view that the new ministry would not last a single parliamentary session was shared by most seasoned observers of the political scene. Its leaders were inexperienced; their main claim to fame being involvement with the Jockey Club, and not administrative expertise. Rockingham also erred in believing that the king would give the new ministry his full

[24] B.L. Add. MSS 57810, fo. 176.

support, and that his ministers would be able to attract members of the old government to their ranks.[25] At first Grenville simply felt flattered that his associates thought him the man to save the country in the event of Rockingham's failure. He was soon compelled to abandon this complacent attitude, and assume the rôle and responsibilities of leader, however, when it became clear Rockingham intended to do more than reverse a few policies of the previous administration.

Indeed, the birth of a Grenville party in opposition as a clearly defined group was a sudden and, for many of his friends, traumatic experience. It arose almost entirely from the purge of office-holders initiated by Rockingham when he took office. In this action few of Grenville's friends or appointees were spared. A slight connection was sufficient to warrant dismissal, and by the autumn of 1765, Grenville found himself nominal leader of a large number of displaced government servants and supporters.[26] There was at this juncture little contact or cohesion within the group. Most of its members were determined simply to seek revenge on the politicians responsible for their dismissal, and they paid little attention to Grenville's reluctance to assume the mantle of leader. There were, nevertheless, four clear sub-divisions within the group as a whole from which Grenville's party would later develop.[27]

The first and easiest to identify, was that made up of eighteen or so MPs who had held office under Grenville, and were later dismissed or tendered resignations in protest at his removal. Several followed Grenville until his death, but not all were his appointees. MPs like Sir Fletcher Norton (MP for Guildford) and Thomas Whately (MP for Ludgershall) came in with Bute and transferred their allegiance in April 1763. They did all have one thing in common, however: Rockingham disliked them and Grenville appeared the best hope of furthering their political careers.

The second group consisted of five politicians indebted to Grenville for past favours. Two MPs, William Hussey (MP for St. Germans) and John Sargent (MP for West Looe),

[25] O'Gorman, *Rise of Party*, pp. 129–31.

[26] A list of changes in office appears in *Fortescue*, i. 129–32.

[27] The analysis that follows is taken from the account in Lawson, Ph.D. thesis, pp. 1–13.

owed their seats to recommendation from Grenville while he was first lord, and a third, Henry Shiffner (MP for Minehead) had been granted a secret service pension.

The third source of support sprang directly from the benevolence of Grenville's rich and influential admirers, who provided about ten followers at Westminster. The most important of these benefactors was Robert Clive. In late 1763 they had made a deal over the East India Company which resulted in Clive's small faction of five MPs being put at Grenville's disposal. It was at the time only a temporary agreement, but, as their friendship grew, Clive bound his adherents firmly to Grenville's cause.

The fourth and largest division within the group was composed of unattached MPs unwilling to support the new administration. It is difficult to put an exact figure on this group because opposition to Rockingham did not always mean support for Grenville. There were about twenty-five independents who had connections with Grenville both in and out of office, and their allegiance is not in doubt. But there were another eighteen voting with Grenville first in government and then in opposition whose motives and opinions remain a mystery. The most interesting of these were a dozen staunch independents like Sir Roger Newdigate (MP for Oxford University) and John Rushout (MP for Evesham). Why these twelve should continue to support Grenville after July 1765 is not altogether clear. True they had all declared their support for his ministry, either through an intermediary or by responding to the circular letter sent out in the winter of 1763-4.[28] But it appears that only one, Robert Waller (MP for Chipping Wycombe), was acquainted personally with Grenville or had any contact with him after July 1765. The answer seems to lie in the unique appeal that Grenville's attitude to parliamentary business held for the so-called Tory country gentlemen. His policies to reduce the unfunded debt, eliminate waste in public expenditure, and, most important of all, force the American colonists to share the expense of their defence were all supported by the country gentlemen because they bore the burden of the land tax. Grenville was

[28] See Tomlinson, M.A. thesis, appendix 'a'.

attentive to their interests while in office, and he reaped the reward of their respect and support when dismissed. James Harris was in no doubt about Grenville's popularity among these members. In a memorandum on the political crisis during May 1765 he noted that there were 'great encomiums of Mr. Grenville by all persons. The persons of the first rank and credit at the Cocoa Tree declared for him. On Tuesday last he had a very full levee, when were Sir James Dashwood, Sir Charles Tynte, Sir Robert Burdett, Sir Walter Bagot, and many others.'[29]

Taken together these various groups gave Grenville a potential voting strength in the coming parliamentary session of approximately seventy-four MPs. This was a remarkable tribute to his standing and prestige at Westminster for he possessed few of the essential attributes needed to secure the loyalty of his followers. His electional influence was negligible. There were no seats at his disposal, and he relied for his own return at Buckingham on the generosity of his brother, Temple. Furthermore, Grenville's brand of opposition politics could hold out no hope of personal reward. His party was overshadowed by the larger Rockinghamite faction at Westminster, and Grenville's voice would carry little weight in negotiations for office. Yet Grenville made light of such handicaps. What he lacked in wealth and influence was compensated for by an unerring personal devotion to the interests of his followers. True, he did not always relish the job of pleading political and electoral favours on behalf of others. In his opinion it was beneath the dignity of a former first lord to take the part of parliamentary whip and election manager. But he neglected few opportunities to defend the fortunes of his followers in either parliament or at the polls. It was in this attention to detail that Grenville's strength as a political leader lay. He did secure a loyal party of followers in the years 1765–70 because his conduct was governed by the welfare of its members. 'Make no mistake,' he told one doubting admirer in May 1768, 'I shall always feel very sensibly any injury done to my friends on my account'.[30]

[29] Cited in Namier and Brooke, *House of Commons*, ii. 543. Sandwich endorsed this view in a letter to Grenville on 24 Nov. 1765, B.L. Add. MSS 57810, fo. 179.

[30] Grenville Letterbook, vol. 2.

Grenville's performance in the Commons sealed these talents as a leader. Opposition proved no hardship to Grenville, whose long parliamentary career included several periods of intense activity at Westminster out of the ministerial fold. Part of his appeal to the seventy or so MPs who saw him as their leader lay in his stature in debate, which no one in the new ministry could rival. Rockingham had ample resources in the Lords but from the day of its inception his administration lacked a natural leader in the Commons. He made several attempts to plug this gap. The recruitment of politicians like Sackville and Nugent, for example, certainly helped in the Commons, but Rockingham never overcame the handicap of piloting cabinet policy through the House without an experienced captain at the helm. What made Grenville's task of securing the loyalty of his large following during 1765–6 that much easier, however, were the policies pursued by Rockingham. When it became clear that Rockingham intended to dismantle Grenville's colonial legislation, the decision angered many MPs who had supported the original measures. There were two main areas of contention. First, if the colonists were excused a share in supporting the troops for their defence then the whole burden would fall on Britain. Second, any retreat on the right of taxation would imply an unacceptable limit on parliamentary sovereignty. Rockingham overcame these objections during the debates on the repeal of the Stamp Act in 1766 by basing his policy on economic expediency, and then insisting on a formal statement of parliament's sovereign rights in the shape of the Declaratory Act.

To the surprise of many observers this strategy worked extremely well; the repeal of the Stamp Act passed through the House with large majorities. Despite Grenville's intense opposition, on what appeared a popular point, he attracted few of the crucial votes from independent members.[31]

[31] It is not intended in this chapter to offer a detailed account of Grenville's campaign against the repeal of the Stamp Act. Two recent studies give an excellent summary: Thomas, *British Politics and the Stamp Act Crisis*, p. 185 *et. seq.* and Langford, *The First Rockingham Administration, passim.* For a broader appraisal of Grenville's attitude to colonial problems in the years 1765–70 see also Lawson, 'George Grenville and America – The Years of Opposition 1765–1770', *W.M.Q.* 37 (1980), 3rd series, 561–76. In this and the following chapter my concern is to make some assessment of Grenville's contribution to party politics through his involvement in the opposition campaigns of the years 1765–70.

He should have learnt his lesson from his days in office, for once it became clear that the king supported the repeal Grenville could only depend on the loyalty of his own followers when the divisions came on. Allied to this Grenville made some fundamental errors in the parliamentary campaign against Rockingham's administration. The session began badly on 17 December 1765 when he attempted to embarrass the government during a debate on the Address, over what he considered its conciliatory attitude to news from America of resistance to the Stamp Act. The House resisted Grenville's suggestion for an amendment expressing disgust at the colonial protest and promising enforcement of obedience to king and parliament, obliging him to withdraw his motion with a severe blow to his pride and prestige.[32] Further errors of judgement were committed during February and March 1766, when the repeal of the Stamp Act and Declaratory Act went through the Commons. Grenville decided to debate these resolutions clause by clause, refuting Rockingham's claim that the Stamp Act had caused severe dislocation of Anglo-American trade and should therefore be repealed. It was not a bad strategy to adopt in the circumstances, for there was no causal connection between the Stamp Act and the slump in the British economy in the mid-1760s. After some skilful performances on the floor of the House, however, this strategy backfired. On 7 February Grenville moved an Address in the Commons calling for the enforcement of all the laws of the kingdom. Many MPs believed this ill-disguised reference to the Stamp Act premature, likely to cause further resistance if not bloodshed. To avoid a direct vote on the Address therefore, the ministry divided the House on a procedural motion and won a resounding victory. Grenville had overplayed his hand and this day marked the end of serious opposition to the Stamp Act. The Rockingham administration built its case for repeal on the evidence of witnesses examined at the Bar of the House, who testified to the decline in the British economy following American protests at the Stamp Act. Grenville in his turn brought forward his own 'experts' intending to show that patterns of trade remained unaffected by his

[32] *Commons Journals*, xxx. 437–8 and Walpole, *Memoirs of George III*, ii. 350–3.

legislation. But his witnesses proved most unreliable, frequently demonstrating through their evidence that the government and not Grenville had the right answers. Grenville, dismayed at this, sought to press even harder and made rash decisions in committees on the repeal, pushing for divisions on a weak point and suffering heavy defeats for his pains. After two sound defeats on 21 February and 4 March, Gilbert Elliot, who had remained in the ministry after Grenville's dismissal, commented that 'Mr. Grenville and his friends are like to have leisure enough to repent of their headstrong and ill-advised conduct'.[33] John Walsh, Clive's agent in London, went further, reporting on 19 February 1766 that Grenville 'does not possess in any eminent degree opposition talents'.[34]

It is a paradox that notwithstanding these errors the campaign did in the long-run help Grenville to maintain his own party intact, simply because he defended those principles and interests closest to the hearts of its members. The failure of the campaign against repeal was a bitter pill to swallow but it did not shake Grenville's confidence, whose attacks during the latter days of Rockingham's ministry were to be more fervent than ever. The government's policies again made this task that much easier, especially a proposed window tax contained in the budget resolutions of 18 April. This measure, as Bamber Gascoyne wrote at the time, provided a perfect opportunity for Grenville to reproach the ministry for its misguided economic policies while voicing the complaints of the country gentlemen against this new tax. On 18 and 21 April 'Mr. Grenville trimmed the Chancellor of the Exchequer beyond imagination', Gascoyne noted, and it came as no surprise to him that in two important divisions on the tax the government majority was severely cut back.[35] Gascoyne was quick to stress that this showed Grenville could still attract the independent votes that had eluded him during the debates on the repeal. Perhaps the most vital factor of all in maintaining Grenville's support during these months, however, was the widely held belief that the ministry could and would not survive. After

[33] *Border Elliots and the Family of Minto*, p. 399.
[34] Clive MSS vol. 52.
[35] Strutt MSS T/B 251/7, vol. 1.

Cumberland's sudden death in October 1765 rumours of a change in personnel were a constant threat to Rockingham's position which he never managed to dispel. A large number of members, in government and opposition alike, considered that the ministers were too inexperienced to shoulder the heavy weight of day-to-day business in the House. Even success in the repeal debates brought no relief. 'We expected', Gilbert Elliot observed in April, 'these holidays might have produced some changes, but I hear of none.'[36] Whether justified or not speculation of this kind strengthened Grenville's hold on his party, and the conviction that he would soon return to office. The king's hatred of his former minister was not as yet common knowledge among his followers, and, if Rockingham fell, it was Grenville that they fully expected to step into the breach.

These assumptions were soon put to the test. In July 1766 Rockingham did leave office, but it was Pitt and not Grenville who was asked to take his place. Though the change was not unforeseen it proved to be a severe blow to Grenville's popularity and standing in the House. Expectation of a quick return to office bound a large group of MPs to Grenville's party, and as this possibility diminished so did the size of his support. It was not a sudden movement. In the first six months of the new administration Pitt, now Chatham, claimed only four of his followers for office: Augustus Hervey, Charles Jenkinson, Hugh Percy, and Hans Stanley. At that stage it was a loss that Grenville could bear, for they were replaced by Lord George Sackville and his two friends, Sir Thomas Pym Hales and Sir John Irwin. Sackville was the only politician proscribed by Chatham from serving in his ministry. He had been a member of Rockingham's administration, but after voting against the repeal of the Stamp Act in February 1766 identified himself with Grenville and his followers at Westminster.[37] He was an effective speaker in the House and later became a member of Grenville's inner circle.

The drift away from Grenville was slow and haphazard. Chatham's incapacity and divisions within the cabinet were

[36] *Border Elliots and the Family of Minto*, p. 399.
[37] Namier and Brooke, *House of Commons*, iii. 393.

to hamper any serious attempt at strengthening the ministry for almost a year. Yet there were signs at the start of the new session in November 1766 that Grenville's popularity as an opposition leader had waned. The first warning note was sounded in early November when he launched an ill-judged campaign against an order in council of 24 September 1766 forbidding the export of grain. Grenville believed the order to be unconstitutional, and had worked hard during the recess contacting friends with a view to supporting a full-scale attack on the issue. There was never any doubt in his mind that the campaign would appeal to all his supporters, especially the country gentlemen, and it was duly launched after the Address on 11 November. Indeed, Grenville was full of optimism for the forthcoming campaign, telling Sir Hew Dalrymple on 24 October

> I hope that there will be a full attendance, and that we shall have the pleasure of seeing you there at the meeting, especially as I hear that some measures are proposed to be immediately taken against the present bounty upon corn, in which the whole landed interest of this kingdom and particularly of Scotland is so essentially concerned.[38]

It brought scant reward. It was not that Grenville had erred in doubting the legality of the order. The government was to admit as much in debate on 18 November. It was simply that he had based his assault on the wrong ground. The majority of MPs accepted the government's argument that, legal or not, the serious disturbances resulting from the shortage of bread during the summer demanded urgent remedial action. They may have supported criticism of the delay in making the order, but Grenville's high constitutional principles hardly seemed relevant at the centre of such a crisis. He had misjudged the mood of the House and it is not surprising that in the only division on the embargo his party suffered a reverse. In a well-attended Committee on the Indemnity Bill on 5 December, the government beat off an adjournment motion by Alexander Wedderburn with a comfortable majority of 118 votes.[39]

[38] Buccleuch MSS GD 110/940/38.

[39] Walpole, *Memoirs of George III*, ii. 286. Little has been written about this campaign since Winstanley's brief but useful account in *Lord Chatham and the Whig Opposition*, pp. 74–5 and pp. 88–9.

A minority of 48 votes could not have fulfilled Grenville's hopes for the campaign. He had certainly experienced some success in the debate: Bamber Gascoyne was moved to comment after the debate on 18 November that 'we may thank Mr. Grenville' for establishing 'the fundamental doctrine to the constitution that the Crown has no power to dispense with the written laws of the realm'.[40] But by acting alone he had lost the initiative and exposed the frailty of his support to the world at large. Over the summer he might have expected to lose a dozen or so supporters but a drop from the mid-seventies to the high forties threatened the chances of all future campaigns. Unfortunately there are no division lists available for Wedderburn's motion on 5 December so it is impossible to know who deserted. It is even in doubt whether the minority consisted of Grenville and Rockingham supporters together or Grenville's followers on their own. On balance it is likely to be the latter. A combined opposition vote of 48 votes would have been a major disaster for both leaders, and there are no post-mortems in their correspondence to suggest such an event took place.[41] Uncertain or not, to avoid a repetition of this sound defeat it was imperative that Grenville and Rockingham concerted their efforts during the next campaign. Grenville himself was not keen on the idea. The bitter struggles over the Stamp Act were still fresh in his mind and he would have preferred to battle on single-handed. Several of his followers, however, adopted a more flexible attitude. They saw the danger in such prejudices, and in early December Grenville's friend Lord Lyttelton made tentative approaches to the Rockinghams with a view of joining forces.[42] The reply from Rockingham was discouraging, but his followers continued the dialogue and on 9 December achieved a minor breakthrough. In the division on Beckford's motion for papers relating to the East India Company, several Rockinghamites openly opposed the government and later voted in the minority with Grenville.[43] All subsequent negotiations were then directed

[40] Strutt MSS T/B 251/7, vol. 1.

[41] Walpole, *Memoirs of George III*, ii. 286 and *Fortescue*, i. 422.

[42] The negotiations are traced in Brooke, *The Chatham Administration*, pp. 81–3.

[43] Walpole, *Memoirs of George III*, ii. 290.

towards persuading their respective leaders to accept the necessity of some sort of union.

It proved to be a difficult task, and Grenville was to suffer a second and more serious reverse in the Commons before he agreed to consider the idea of co-operation. The point at issue was the government's attitude to colonial taxation. Grenville doubted the cabinet's resolve on this thorny issue and in the new year decided to bring the matter to a head. The trap was sprung on 26 January 1767 in the Committee of Supply on the army estimates when Grenville moved a surprise amendment proposing that 'the Colonies should pay the regiments employed there'.[44] It was a thin House and Grenville had made no plans for the debate, but it did not deter him from forcing a division. After a short discussion with speeches of support from Lord George Sackville and Thomas Pitt, the Committee divided on the amendment. The result proved very unsatisfactory, with the government winning a comfortable majority by 106 to 35 votes. This represented a further significant drop in Grenville's numerical support, but he was unwilling to heed the warning carried by these figures. The following day the report from the Committee was read to the House and Grenville made another attempt to win approval for his amendment. It was a débâcle, for this time he could muster only nineteen votes.[45] Though undoubtedly disturbed at the result, Grenville played down the importance of these divisions. In his wife's diary there is nothing but a bland note expressing disappointment at the low number of MPs present in the House. 'The Army Extraordinaries for England and America were passed', runs the account, 'with no more than about 150 Members present.'[46] It was simply a brave attempt to cover a most unpalatable truth. A poor attendance notwithstanding, he must have been aware that any remaining hope of leading a large party back into office had now come to an end.

It is not difficult to find the immediate causes behind this eclipse. The debates did take place in a thin House, and it is clear from Grenville's remarks in his correspondence that a

[44] Walpole, *Memoirs of George III*, ii. 293.
[45] *Commons Journals*, xxxi. 76.
[46] *Grenville Papers*, iv. 211.

large number of his supporters had not bothered to attend the Committee. It is also apparent that, unlike the campaign against the corn embargo, Grenville had had no option but to act alone in this matter. On America, the Rockinghams to a man had refused to contemplate any form of co-operation, and the Bedfords, as Walpole put it, were 'being kept back by the Duchess, still restless to return to court'.[47] But these factors alone do not explain the wholesale decimation of Grenville's following since Rockingham's fall. This had begun before Christmas when the ministry had enjoyed unexpected successes in the debates on the corn embargo. Rather than wither under Grenville's attacks, Chatham's administration had stood firm, and, with the king's full support, circumstances suggested it would continue to do so for the foreseeable future. Among the less stout-hearted of Grenville's followers, unaccustomed to long periods of opposition, such resolution was greeted with dismay. If Chatham remained in office they would have no choice but to mount another fruitless campaign in opposition, and this prospect was to cost Grenville the support of at least seven former associates.

Grenville's cause in these months had not been helped by the eagerness of the Bedfords to re-enter office. Their refusal to support him in January while negotiations were underway merely reinforced the impression that the ministry was here to stay. Nevertheless Grenville's conduct in the House during the Committee of Supply had done little to counter the growing disillusionment among his supporters. If the choice of a campaign on the corn embargo was unfortunate, the unheralded attack during the army estimates was disastrous. It not only quashed any hope of co-operation with Rockingham and Bedford but also failed to attract the crucial support of independent MPs. Grenville did not expect the set-back, and apart from low attendances in the Chamber proffered no other reason to excuse it. Yet the clues to his defeat were evident in the debates themselves and Grenville simply overlooked them. He makes no mention in his correspondence, for example, of Townshend's much-discussed promise on 26 January to introduce an American tax later that session.

[47] *Memoirs of George III*, ii. 294.

This answered Grenville's demand for an assurance on colonial policy, and no independent rose to support the amendment after Townshend had delivered the pledge. True they did not know the exact form of the intended tax, but to continue the attack on similar ground the following day simply appeared as opposition for its own sake. This was an unpardonable error for a politician of Grenville's experience. On controversial issues there was, as Brooke points out, 'a tendency on the part of the independents to give the government the benefit of the doubt'.[48] If, as Townshend maintained, the government was committed to raising an American revenue then it was only fair to allow such measures to come on before passing judgement.

Grenville did not make the same mistake again. The minority of 27 January caused sufficient embarrassment to ensure that there were no further lone campaigns that session. Over the next six weeks both Grenville and Rockinghm realized that if they were to make any impression on the government's majority in the House past differences would have to be buried. This did not repair the lasting damage inflicted on Grenville's party during the winter months. The debates in January had shown conclusively that he could no longer rely on the support of the unattached members, nor even depend on a large number of those who had served in his ministry. Indeed, accounting for absentees in these debates, it was clear that no more than thirty-five MPs were now prepared to support the man irrespective of the measures at hand. Yet co-operation during the remaining four months of the session did protect his following from total eclipse. The government's inquiry into the East India Company, in particular, provided Grenville with an unexpected opportunity to recover some ground lost in January. The inquiry was unpopular in the House, and there were no ideological differences with Rockingham to hinder an all-out campaign against the plan to extract revenue from the Company.[49] Battles were fought on the carefully chosen ground surrounding the sanctity of charters and private property, and Grenville's defence of these cherished maxims aroused the

[48] Namier and Brooke, *House of Commons*, i. 197.

[49] For further details of this campaign see Lawson, '*Parliament and the First East India Inquiry, 1767*', *P.H.Y.B.* I (1983).

sympathy of MPs on all sides of the House. Not surprisingly the government repeatedly denied the claim that its inquiry was proceeding along these lines, but majorities of 33 and 14 bear witness to the potency of the opposition attack.[50] Whatever the motive for the inquiry Grenville certainly had cause to be grateful for its existence. The debates and divisions in committee not only revived his sagging popularity but, more significant, delayed the defection of several important figures, such as Sir Fletcher Norton, from his party.

He was assisted in this brief resurgence by two exceptional strokes of good fortune. The first was the government's inability to solve its own internal divisions in these months. The East India Company inquiry was Chatham's pet scheme, but without his guiding hand the cabinet was soon at loggerheads over what direction the policy should take. Worse still, the argument spilt over into the Commons. In the East India Committee, Conway and Townshend frequently appeared as passengers in the administration, and made no secret of the fact that they were carrying out a policy in which they had no confidence. It was an open invitation for a sustained opposition attack and both Grenville and the Rockinghams exploited it to the full. The second occurred on 27 February when the opposition won a surprise victory in the Commons on a proposal to reduce the land tax from 4*s*. to 3*s*. in the pound. The motion for a reduction had been well planned. Whately floated the idea to Grenville in October 1766 as a 'popular topic . . . to make a stand upon', and the Rockinghams were in full agreement with the idea from the beginning.[51] They saw its appeal to the country gentlemen and began earnest preparations for the debate in early February. On the day expectations were high, but neither Grenville nor Rockingham actually foresaw a victory on the floor of the House. In fact had the government made even the slightest effort to whip in support there seems little doubt that the amendment would have been defeated. The king certainly recognized this fact, and spelt out in a letter to Conway on the same day the necessity of proceeding with ministerial business as before: 'I doubt not on all

[50] These figures refer to the debates on the Company of 9 March and 7 April 1767 respectively, *Ryder Diary*, p. 335 and p. 338.

[51] *Grenville Papers*, iii. 336.

other occasions', he wrote, 'a great majority will appear in their favour.'[52] Yet this did not detract from the personal acclaim Grenville received after the division. Though Dowdeswell for the Rockinghams moved the amendment, it was Grenville who took the credit for the government's defeat. 'The joy in the House of Commons was very great', noted Mrs Grenville in her diary, 'all the county gentlemen coming round Mr. Grenville, shaking him by the hand, and testifying the greatest satisfaction.'[53] In view of Dowdeswell's role in the debate, Rockingham could rightly feel aggrieved at the outcome. An analysis of the majority, however, reveals the source of Grenville's euphoria.[54] In the division no less than fifty-nine of the seventy-four MPs he had carried into opposition voted in the majority that day. Moreover, of the fifteen lost votes only three, Jenkinson, Percy, and Stanley supported the government. The remaining eleven were either absent, like Clive and Wedderburn, or simply abstained, as did Augustus Hervey, when the division was called. For many of the sixty-one MPs there would be no further votes in opposition, and Grenville did not deceive himself into believing that the government was on the run. But in the light of January's set-back, there is no doubt that the victory provided a much needed boost to morale. There had been no squabbles over principle to bedevil preparations for the debate, and the result had confirmed Whately's claim that a united opposition could make an impression on the government's majority in the Commons.

The impact of this unexpected victory, in the midst of the successful East India campaign, was soon felt in the negotiations for an opposition union. Both Grenville and Rockingham came to believe that Chatham's ministry could be toppled on the floor of the House, and in late March began earnest discussion on posts in a new ministry. They made little headway. Each office was disputed, and they could find no common ground beyond the need to continue the parliamentary attack. Nevertheless it proved to be a significant step in Grenville's opposition career. He was able for the first time to play the rôle of party leader outside the

[52] *Fortescue*, i. 454.
[53] *Grenville Papers*, iv. 212. See also Walpole, *Memoirs of George III*, ii. 297–9.
[54] This appears in B.L. Add. MSS 33002, fos. 470–3 and 33037, fos. 377–89.

House, and there is no doubt that he relished the challenge and responsibility. All the negotiations were handled by Grenville personally, and distribution of offices in the projected administration remained firmly under his control. In view of his inexperience in this art of opposition politics, he managed the task well. Though he could be criticized for refusing to meet Rockingham face to face, their mutual antipathy probably made this a wise decision. Of the two Grenville certainly had the more realistic approach to future policy. He sought to build on the success of the East India campaign and assured Rockingham that on no account would he recommend the reintroduction of the Stamp Act.[55] Grenville could have made no greater concession, and in return he demanded the highest office of first lord for himself. At this point deadlock ensued, and Grenville put the blame wholly at Rockingham's door. He believed, and not without foundation, that his party possessed the best speakers in the Commons, and should be allowed to govern matters there while Rockingham orchestrated his followers in the Lords.[56] Grenville may have lacked electoral influence and a large following at Westminster but he was not prepared to accept the junior partnership allotted by Rockingham in the planned ministry. It was a bold stand and several of Rockingham's friends were to bemoan his obduracy. But there were no further concessions. Grenville was negotiating on the strength of his successes in the House in February and March, and behaved like a party leader expecting a quick return to office.

The final breakdown of the negotiation on 8 April passed without any hint of remorse on Grenville's part. Formal contact had certainly failed but Grenville saw no reason to abandon the parliamentary campaign which still offered hope of a united opposition front. During the remaining weeks of the session his attacks on the government's imperial policies continued unabated, and he made every effort to avoid giving direct offence to the other opposition groups. Unfortunately Rockingham had less heart for the battle than Grenville and the breakthrough promised in the debates

[55] The message was carried by Bedfordite, Richard Rigby, to the Duke of Newcastle on 1 April 1767, B.L. Add. MSS 32981, fos. 1–3.

[56] *Grenville Papers*, iv. 220.

of 27 February and 9 March never materialized. There were no open disagreements between the two, but their failure to establish a consistent approach to the government's American legislation bedevilled the whole opposition campaign. More depressing still, when a limited agreement on strategy was finally attained, the opposition suffered a sound and unexpected defeat. On 13 May all three opposition parties launched an attack on Charles Townshend's proposals for dealing with colonial resistance to parliamentary law. Grenville himself led the attack and put forward the amendment on which the division took place.[57] To judge from the result it made little impression on the House. Despite all its guns firing in unison the opposition mustered only ninety-eight votes.[58] The outcome was sufficient to persuade several Rockinghamites that there was nothing to be gained from continuing the campaign. 'Our friends were a good deal surprised at the division' James West told Newcastle, 'Sir W. Meredith and Mr. Aufrere told me they could come no more to Parliament this year and most people look upon the session as in a manner over and are going out of town.'[59] In fact when the Committee on American Papers reported the resolutions on New York two days later, 'Lord Rockingham's friends', Ryder noted, voted with the government in the two divisions that took place.[60] This did not lessen Grenville's resolve to attack the government's policies at every opportunity, but without Rockingham's support he made little progress. As the end of the session in June drew near, attendances fell and he was unable to capture the crucial independent vote.

Grenville mused long and hard over the lost opportunities of these weeks, but never regretted the energy expended during the campaign. His standing at Westminster could not have been higher, and in late May it received a further boost with the publication of the *Political Register*; a monthly magazine that lent its full support to Grenville and his

[57] *Ryder Diary*, pp. 342–7.

[58] The full result was 180 to 98 votes. In mitigation, James Harris pleaded that a miscount of the opposition benches had taken place, Thomas, *British Politics and the Stamp Act Crisis*, p. 325.

[59] B.L. Add. MSS 32981, fos. 379–80.

[60] *Diary*, p. 348.

followers in parliament. Its origins are obscure. It was begun by Temple's friend, John Almon, primarily as the result of promised contributions from John Wilkes and other political celebrities. But its appearance in May 1767 after the encouraging opposition campaign of the spring cannot be wholly coincidental. Indeed, the first two issues are dominated by accounts of the land tax debate and discussions on the East India Company's relations with the ministry.[61] No less uncertain is Grenville's involvement with the journal. The problem lies in the fact that his attitude to the press throughout his years of opposition was neither rational nor consistent. On the one hand he disliked being attacked in the press. In an exchange with Conway on 24 November 1767, for example he complained bitterly that he 'had been much misrepresented in libels', and two years later wrote in similar vein that he opposed publication of his speeches as 'it will cast down upon me a heap of scurrilous abuse from the ministerial writers'.[62] Yet, on the other, he did not hesitate to use the press for his own political ends. In office he had enjoyed the services of three talented propagandists, Charles Lloyd, Tobias Smollet, and Thomas Whately, and after July 1765 was fortunate enough to carry two of them, Whately and Lloyd, into opposition. They continued to write pamphlets and contributions to the press supporting his line in parliament, and in 1768 were joined by William Knox, a long-standing admirer of Grenville with expert knowledge of colonial affairs. Grenville never expressed concern at their work, and in the case of Knox actually collaborated in the preparation of his two most famous pamphlets and certain pieces for the press.[63] The only condition ever stipulated was that the work should make no mention of Grenville's name or imply his formal approval. Whether this anonymity was prompted by a fear of becoming involved in a heated press war or simply a belief that it was beneath

[61] There were fifteen issues of the journal under Almon's ownership, the last appearing in July 1768. They were then collected and published by Almon in two volumes, and all references that follow refer to these bound volumes.

[62] Walpole, *Memoirs of George III*, iv. 82.

[63] The tracts appearing in 1768 and 1769 were *The Present State of the Nation* and *The Controversey Between Great Britain and the Colonies Reviewed*; for a discussion on a piece for the press see Knox's letter to Grenville on 20 Aug. 1768, *Grenville Papers*, iv. 346–7.

the dignity of a former first lord to voice his opinions in the popular newspapers is not clear. But there is no doubt that it smacked of hypocrisy, and an incident following the December 1767 issue of the *Political Register* illustrates the point. In this edition a strongly worded article had appeared, attacking the Duke of Bedford's decision to lead his faction back into office. It was not signed but the piece clearly bore Grenville's stamp. The cryptic style was his, and it contained detailed information of conversations with Bedford which could only have come from Grenville himself.[64] Nevertheless he denied any responsibility for the article. 'I took occasion', he told Whately after hearing of the furore it had caused, 'to assure some of the Duke of Bedford's friends of this before I left London, both on my brother's part as well as my own.'[65] Bedford, quite justifiably, was not convinced and never forgot the incident. Articles appeared in the *Political Register* both before and after this incident to suggest that Grenville's denial was one of convenience.

Whatever the degree of complicity by Grenville, his followers certainly reaped considerable benefit from the wider circulation of their opposition campaign. It was, as Almon told Wilkes, an extremely popular and respected magazine. 'The *Political Register*', he reported during the summer, 'succeeds beyond my most sanguine expectations. It is become the fashionable publication of the times. All parties buy it, and the public approve it.'[66] Its importance was twofold. First it published accurate and detailed accounts of debates in the House and Leadenhall Street, and second it provided a unique platform for the Grenvillite view on more fundamental political issues. Paramount amongst the latter was a vindictive obsession with the friendship between the king and the Earl of Bute. It was a blind phobia based on Grenville's experiences in office. No matter what the evidence to the contrary Grenville believed that Bute still exerted some mysterious political influence over the king. This 'blast of Scottish tyranny', as one correspondent put it,

[64] The piece was published under the heading, 'A word of Parting to his Grace the D[uke] of B[edford]', Almon, *Political Register*, i. 465–8.

[65] *Grenville Papers*, iv. 203–4 (29 Dec. 1767).

[66] Cited in Rea, *The English Press in Politics*, pp. 134–5.

pervaded every constitutional essay that appeared in the magazine, and was always accompanied by some of the choicest invective. Though the Rockinghams were later to claim this shibboleth for their own, Bute's mythological 'power behind the throne' had entered Grenville's imagination long before they took office. It had soured his relations with the king when first lord, especially in May 1765 when Grenville demanded a formal undertaking that Bute would be excluded from politics. His closest friends were aware that Grenville's animosity towards Bute had not waned in his two years of opposition, and neglected no opportunity to reinforce his misplaced suspicions about the activities of 'the Scotch Mouse'.[67] The articles that appeared in the *Political Register* used more forthright terms to attack Bute, but, at root, they were simply public statements of Grenville's private prejudices.

Two other recurring themes in the magazine were also close to Grenville's heart; excessive public expenditure and the increased national debt. Articles on these topics amplified the attacks made by Grenville in the House. In March 1768, for example, Almon published a long piece outlining a scheme to pay off the national debt by restoring trade with the colonies and reducing government expenses. It contained ideas expressed many times before by Grenville, and, significantly, they had been raised only three weeks earlier in the budget debate of 8 February.[68] In fact the only surprising factor about the magazine's content is the infrequency of articles on the American colonies. It has been maintained that in 1767 advocacy of parliamentary sovereignty over the colonies was the only ideological issue binding Grenville's party together.[69] The prevalent views expressed in the *Political Register*, however, do not justify such an assertion. Antipathy towards Bute and a desire to reduce the burden of taxation were of equal if not greater importance to ensuring continued support for Grenville's opposition campaign. Indeed, his self-appointed watch on the public purse consistently attracted votes from all sides

[67] This was Lyttelton's description to Grenville on 27 July 1767, *Grenville Papers*, iv. 114.
[68] H.M.C. Foljambe, p. 239.
[69] Brooke, *The Chatham Administration*, p. 263.

of the House and it would have been a gross error of judgement not to exploit this line of attack. Sir Roger Newdigate was not alone when he told MPs on 24 March 1767 that Grenville's stand on this principle made him the only minister for which he felt admiration.[70]

Almon's decision to sell the journal in June 1768 was a severe blow to Grenville's party. It was not that all outlets to the press were cut off; until his death in November 1770, Knox, Lloyd, and Whately continued to write articles and pamphlets in support of Grenville's attack on government policy. Furthermore, Grenville received unexpected backing from the pen of Junius. Though he certainly had no idea of the author's identity, Grenville was, from the very beginning, Junius's most favoured politician. The tone was set in the first published letter signed by Junius of January 1769. In a long polemic on the record of each administration since the accession of George III, Grenville was the only minister to escape censure. His trade and colonial policies, Junius wrote, were the sole means 'of giving any sensible relief to foreign trade and to the weight of the public debt'.[71] Over the next two years the praise for other policies was similarly unreserved. To the campaigns on America, the East India Company and Wilkes, Junius gave unstinting support, fulfilling to the limit his private promise not to rest 'Untill you are Minister'.[72] It seems fair to assume that Grenville welcomed this unsolicited praise, but to what extent it strengthened his hand in parliament is uncertain.[73] Junius undoubtedly played an important rôle in weakening Grafton's resolve to govern, but he fell far short of his aim to carry Grenville back to power. Grafton's successor was the wily North who soundly defeated the opposition's hopes of bringing the government down on the floor of the House. Hard though he strove, Junius never surpassed Almon's achievement in the years 1767–8. The loss of the *Political Register* was felt simply because it represented the official

[70] Walpole, *Memoirs of George III*, ii. 312.

[71] *Public Advertizer*, 21 Jan. 1769, cited in Cannon, *The Letters of Junius*, p. 29.

[72] Ibid., p. 449 (20 Oct. 1768).

[73] Grenville certainly did not complain when Junius, writing under the pseudonym of 'C', pledged himself to 'your cause and to *you* alone' on 30 Sept. 1768, *Grenville Papers*, iv. 355.

voice of Grenville's party. It published and reflected Grenville's opinions accurately, and, most important of all, related them directly to the parliamentary campaign.

The sole comfort to be gleaned from its sale was the fact that it had supported Grenville through the most difficult nine months of his brief period in opposition. The failure to capitalize on the parliamentary campaign of the spring was a disappointment, but it proved to be only the first in a series of set-backs that disrupted Grenville's party in late 1767 and early 1768. The trouble began in July 1767 when the Duke of Grafton opened negotiations with the opposition with a view to rebuilding the administration. The move took no one by surprise. Conway, the leader in the Commons, had been threatening resignation since late May, and it seemed a perfect opportunity for Grenville to reassert his claim for the post of first lord. To the obvious astonishment of many of his followers, however, Grenville refused to press for the office, maintaining an air of passive disinterest throughout the negotiations. In contrast to the discussions with Rockingham during March, he did not take any direct part in the three weeks of intensive talks. Messages to and from the other opposition leaders were carried through Richard Rigby, a follower of Bedford, and, more remarkable still, Grenville disclaimed all pretensions to serve in the new ministry. His only conditions for supporting this future government were a firm colonial policy and adequate reward, in terms of offices, for his friends.[74]

Many reasons have been put forward to explain his behaviour during these weeks. It is argued by Brooke that Grenville knew the king had vetoed his return to office and was simply adopting a realistic attitude to his exclusion.[75] This certainly provides the key to his rapid change of heart since the negotiations of the spring when Grenville was eager and willing to take office. But it is not borne out by the evidence available. Grenville never admits to any veto in his private papers and there is a note in Grafton's memoir for 4 July 1767 that states: 'The idea of Lord Temple or Mr. Grenville in a great office was no ways a hindrance in

[74] The fullest account of the negotiations in the summer of 1767 is still that by Brooke, *The Chatham Administration*, pp. 162–217.

[75] Ibid., pp. 264–5.

the king's mind; if by it, I could remain in my post . . .'[76] The assumption that a veto existed is of modern invention and it was not one shared by Grenville's contemporaries. Long after this incident, rumours of Grenville's return to office were being taken seriously. In the summer of 1768, for example, Benjamin Franklin was in no doubt that Grenville would be brought in to strengthen the ministry. Only his experience of colonial affairs and standing in the Commons, he intimated to Joseph Galloway on 12 July, would compensate for Shelburne's imminent departure as southern secretary. 'Several of the Bedford party being new got in', Franklin wrote, 'it has been for some time apprehended that they would sooner or later draw their friend Mr. Grenville in after them.'[77] The following year Robert Clive was equally confident of Grenville's return to office. He had drawn up a plan for the future of the East India Company and commented to Claud Russell on 19 February 1769 'if Lord Chatham and Mr. Grenville should appear once more at the head of affairs – of which there is some prospect – they are the only men capable in my opinion of embracing such ideas which you knew are extensive'.[78] Even after Grafton's fall from office in January 1770, Grenville was being discussed as a successor. A note in Sir Gilbert Elliot's diary reads: 'Wished Grenville could have been got, in which North agreed, but said it was impossible.'[79] In fact North's objections were related to Grenville in the Commons on 31 January when Elliot told him that 'had he not entered into factious combinations, he knew Grenville would have been entreated to save his country'.[80] This statement did not imply permanent exclusion, simply that it was politically inexpedient to bring Grenville in at that juncture. North himself certainly grieved his absence, describing him as 'the fittest man of all for the office held'.[81]

A more likely explanation of Grenville's diffidence during July 1767 lies with his repeated warnings that Grafton's

[76] *Grafton Autobiography*, p. 151.
[77] *Franklin Papers*, xv. 164.
[78] Clive MSS vol. 62.
[79] *Border Elliots and the Family of Minto*, p. 407.
[80] Walpole, *Memoirs of George III*, iv. 77.
[81] From a memo by James Harris cited in Namier and Brooke, *House of Commons*, iii. 207.

overtures should not be taken at face value. He was suspicious of his motives from the start, and it clearly influenced his decision to take a back seat in the negotiation. There was no doubt in his mind, as he told Suffolk after a week of talks, that the ministry was merely playing one opposition group off against the other in the hope of destroying such little trust as existed between them.[82] Grenville was particularly angry at Grafton's dealings with Rockingham, and also about the dispute over whether the opposition leader had been granted official permission to form a new administration. 'I have been convinced from the beginning', he told Temple on 16 July, 'that Lord Rockingham overrated the powers supposed to be given him . . . and that the Duke of Grafton acted exactly the same part with Lord Gower and Lord Rockingham, with the same view of disuniting and getting some individuals.'[83] In retrospect Grenville probably credited Grafton with more guile than he possessed, but his belief that the discussion would prove abortive was quite correct. On 18 July he wrote to Temple expressing no hope for a successful end to the negotiation. 'I sincerely wish', he observed, 'that we should be together when the dénouement of this farce appears, which I think is not far off.'[84] The following week all official contact between government and opposition came to an end.

The accuracy of the prediction was of no satisfaction to Grenville, for the breakdown in negotiations fulfilled his worst fears. Whether deliberately or not Grafton had driven the opposition into disunion and emerged with a much stronger hand to play in cabinet. The government had not crushed the idea of a united front, but the seeds of distrust had been sown deep enough to render any further attempts to bring Bedford, Grenville, and Rockingham together in vain. It was a severe reverse for Grenville, but his role as a bystander had done little to prevent the outcome. If the strength of his suspicions about Grafton's motives had been matched by direct action, the negotiation may well have taken a different course. In all other respects he had played a model part as leader. He laid down terms for entry into

[82] Grenville to Suffolk, 12 July 1767, Grenville Letterbook, vol. 2.
[83] *Grenville Papers*, iv. 53.
[84] Ibid., p. 60.

the administration and informed his followers of developments throughout the talks. Indeed, it was in these weeks that a rudimentary organization for disseminating information emerged within Grenville's party.[85] The structure was crude and based on three ill-defined layers of command. At the top were Grenville's most influential supporters: Buckinghamshire, Clive, Lord George Sackville, and Suffolk. They provided Grenville with the greatest number of followers, and he sent personal dispatches to each one explaining the basis of the negotiation and his response to Grafton's offer. In part this was simply a matter of courtesy, but, as he told Clive on 23 July, they were also expected to relay the intelligence to their own associates within the party. The second tier was controlled by Grenville's independent friends of long-standing: Thomas Grosvenor and Edward Kynaston. Their link with unattached MPs at Westminster was vital to Grenville's support, and he made quite sure that they were kept abreast of significant events during the negotiation. Of the two Grosvenor was the more favoured. He visited Grenville at his country seat, Wotton, in August and supported him in opposition until his death. The third and most junior command was in the hands of Grenville's 'man of business', Thomas Whately. It was his job to brief members of the party that Grenville had not contacted or seen personally over the months of July and August. In practice this entailed no more than calling at the homes of supporters, like Sir Fletcher Norton and Alexander Wedderburn in London, but Grenville attached great importance to the task and gave strict instructions that it should be carried out diligently.[86] Crude though it was, there is no doubt that the system proved reliable. By September only those members on the fringe of the party were ignorant of the negotiation and its outcome. Moreover, Grenville's lieutenants had certainly performed their duties efficiently. His behaviour during the discussions was universally approved, and if there were reservations about the decision to forgo office none were voiced. It was two

[85] All references for the account of the party that follow, unless otherwise stated, are taken from entries in the Grenville Letterbook, vol. 2 for July, Aug., and Sept. 1767.

[86] *Grenville Papers*, iv. 64–5.

years before these lines of communication were reopened but the experience of these weeks was not forgotten. The basic organization remained intact, and it was to play a crucial part in the successful opposition campaign on Wilkes and the Middlesex election.

The loyalty of his followers during the first two years of opposition and these negotiations, however, could not conceal the weakness of Grenville's position at their close in August 1767. His conduct had left the party isolated. The opportunity of leading his supporters back into office had now vanished for the foreseeable future, and the opposition alliance was in ruins. Grenville demonstrated little active concern at this unpromising outlook, preferring to spend the summer at Wotton resting and visiting friends. His confidence remained high because he did not consider the situation beyond redemption. Grenville admitted to Charles Lloyd that he had enjoyed the campaign since 1765 just as much as his time in office, and he believed the opposition would go from strength to strength.[87] Furthermore, Grenville obviously basked in the reputation of elder statesman that he had begun to assume after the victory on the land tax, and did not see why similar victories could not be achieved in the future. Far from being despondent after the failure of the talks in the summer, therefore, Grenville looked forward to the coming parliamentary session full of optimism; quite unaware that the united opposition front of Bedford, Grenville, and Rockingham was soon to be just a fond memory.

[87] B.L. Add. MSS 57818, fo. 113.

Chapter VIII

The Indian Summer, 1767–1770

In his account of Grenville's career after 1767, Sir Lewis Namier presents the view of a man detached from the struggles of office, leaving the day-to-day running of his party in the Commons to his chief lieutenant, Thomas Whately. The urgency to prove that he was still a potent force at Westminster disappeared, and Grenville came to enjoy a position 'as foremost senior statesman among commoners'.[1] It is a generous portrait of the last days of a politician for whom the author had little sympathy, but it requires severe revision. In the years 1767–70, Grenville neither put himself above the practical obligations of party leadership nor modified his parliamentary attacks on the government of the day. His involvement in electioneering during the 1768 general election, the campaigns on the Middlesex election of 1769, and his personal victory in reforming the trials of controverted elections in 1770 bear witness to a politician still interested in occupying the highest offices of state. Grenville's commitment to the opposition cause never lapsed or waned. In the autumn of 1769, when questioned by Whately about the prospects of defeating the Grafton administration on the floor of the House before returning to office, Grenville replied bluntly that it would be 'an object worth contending for'.[2] He did not attend Westminster for the sake of appearances or simply to relax in the role of elder statesman. For activity and achievement, these three years represented Grenville's halcyon days in opposition.

Optimism and enthusiasm amongst Grenville's partners in opposition was in short supply, however, as the 1767–8 parliamentary session approached. Grenville first sensed trouble in early November when he contacted Bedford with

[1] *House of Commons*, ii. 544.
[2] Grenville Letterbook, vol. 2 (26 Nov. 1769).

a view to making plans for the opening debates. The tone of the duke's letters showed that relations between them had cooled considerably since the summer. After the events of July, Bedford had grown tired of the search for opposition union. He met Grenville on 21 November, three days before the session opened, but refused to commit himself on policies for the opposition attack. Grenville did not suspect anything was amiss, and they agreed to take their own line during the initial debates in the hope of building on parliament's reaction to the Address. It proved a vain hope. Since the beginning of the month Bedford had been considering overtures from the ministry about bringing his followers into government. The only problem he faced on acceptance was awaiting the right opportunity, and when parliament met on 24 November Grenville himself provided it.

It was a thin House and the opening exchanges on the Address gave little indication that it was to be an auspicious occasion. Dowdeswell did move an amendment requesting the government find ways of 'assisting the manufactures of his Majesty's kingdom, and preserving, extending, and improving its foreign trade', but it was nothing more than token resistance.[3] After speeches of support by Burke and Wedderburn, it was strongly opposed by the ministry and the debate fizzled out without a division being forced. The only indication that something might be amiss was the fact that Grenville uttered 'not a syllable' during these proceedings.[4] Whether or not he thought the amendment futile is not certain, but this self-inflicted silence was soon abandoned when a general conversation developed on issues raised in the previous speeches. It would seem from the scant information available that two matters in particular had caught his attention. One was a veiled accusation by Conway that Grenville had mishandled negotiations with Spain while in office – an argument that they were apparently able to settle amicably on the floor of the House. The other concerned a brief and unspecified reference to America by Nathaniel Ryder, the seconder of the Address, upon which Dowdeswell made some comment.[5] Though the actual words

[3] *Commons Journals*, xxxi. 422.
[4] *Burke Correspondence*, i. 336.
[5] Ryder MSS vol. 434, doc. 46.

Dowdeswell spoke are unknown, they were sufficient to prompt Grenville into warning the House against impending disaster in the colonies. He had recently received an account from the *Boston Gazette* 'inciting the people to rebel', he declared, on which no action at all had been taken: indeed, he added bitterly, 'the governor there had no power to punish the printer'.[6] The violence of this attack clearly took MPs by surprise, but it did not end there. Turning to the Rockinghams, and Dowdeswell in particular, he launched an equally virulent assault on their attitude to America. Unfortunately the full text of his speech was not recorded. After repeating 'the necessity of enforcing the superiority of this country over the colonies', it appears that he turned to Dowdeswell and told him quite bluntly, 'that there were persons of contrary sentiments whom he never would support in power or co-operate with; and that he would hold the same distance from them that he would from those who opposed the principles of the revolution.'[7]

All hopes of a united opposition front ended with this speech. Grenville knew this and to judge from the brief note in his wife's diary, felt no remorse. 'Mr. Grenville, in the speech he made yesterday at the opening of the session of Parliament, upon the general state of things', she wrote, 'took occasion to make such declarations upon his American ideas as plainly showed that he was upon very different ground from the Rockinghams.'[8] That he should have reacted in this way is not altogether surprising. During negotiations of the previous nine months to unite the three opposition groups, Grenville had become increasingly disillusioned with Rockingham's behaviour. It began in March and April when the two groups discussed places and policies in any future ministry. Apart from the treasury, Rockingham had insisted as *sine qua non* that Grenville 'should have nothing to do with North America'.[9] This had upset Grenville and, though he gave assurances that 'nothing would be attempted or proposed' contrary to Rockingham's wishes, he firmly refused to concede the treasury.[10] He did not want

[6] Walpole, *Memoirs of George III*, iii. 84.
[7] B.L. Add. MSS 32987, fo. 113 and W.W.M. R1–891.
[8] *Grenville Papers*, iv. 235.
[9] Memo by the Duke of Newcastle, B.L. Add. MSS 32980, fo. 450.
[10] B.L. Add. MSS 32981, fo. 34 (Newcastle to Rockingham, 4 Apr. 1767).

to enter a cabinet in which policy was dictated by Rockingham and his friends, and as he had given ground in the negotiations he fully expected Rockingham to do the same. Rockingham refused; he wanted total control, and the negotiations broke down with Grenville suspecting that no sacrifice on his part would be high enough for his opposition 'friends'. This feeling was further enhanced by the negotiations of July 1767. In these discussions Grenville did actually give up all personal pretensions to office, and asked only for the promotion of his followers. Though he did request an assurance on colonial policy by way of a rider, these concessions on places were, in his opinion, not reciprocated by Rockingham.[11] Indeed the only messages that Grenville received in late July were rumours of Rockingham telling the king that he and Temple were responsible for the failure of the negotiations, and of the anger that he had expressed at Grenville's demand for a guarantee on colonial policy. The basis of mutual understanding required for negotiations of this kind, therefore, was never established, and long before November it must have been clear to Grenville that, in Brooke's words, Rockingham 'resented Grenville's interference at all points'.[12]

This is not to say that the attack on Dowdeswell of 24 November was in any way planned. Despite the obvious temptation to embarrass Rockingham, Grenville tactfully kept his disagreements and disappointments to himself during these months. Even in the days immediately before the debate on the Address it seems clear that no overall strategy had been settled. As late as 12 November Grenville was telling one correspondent that nothing 'of much moment except the Common Supplies is intended before the holidays'.[13] In fact it would seem that the precursor of the assault was nothing more than a piece of idle gossip from Lord Lyttelton. On 23 November Grenville held a dinner party for his friends, and during the evening Lyttelton happened to mention some political information gleaned from a recent conversation with Bedford. Whether he intended to shock Grenville is not known, but the news he had to tell was certainly provocative. 'The Duke of Bedford

[11] *Grenville Papers*, iv. 59.
[12] *The Chatham Administration*, p. 208.
[13] Grenville Letterbook, vol. 2.

told him [Lyttelton]', runs Mrs Grenville's account of the story, 'that he had heard from the Duke of Bridgewater that Lord Rockingham had declared that he would never be of any administration in which any Grenville was to have a part.' Though Lyttelton did do Grenville the honour of adding Bedford's 'great indignation' at these remarks by way of a conclusion, nothing could have repaired the damage caused by the initial statement.[14] Grenville had forgiven Rockingham much over the past nine months but this personalized attack obviously went too far. He was both angry and insulted when he attended the House the next day, and it took little provocation on Dowdeswell's part for Grenville to deliver the impromptu declaration of principle. Rockingham himself did not deny the gist of Lyttelton's report, and perhaps the only mistake that Grenville made was in assuming that he spoke for all his followers.[15] Hardwicke, for example, was very annoyed with his leader's ill-timed comments, and later bemoaned 'the extravagance of Lord Rockingham's pretensions' to Lyttelton himself.[16] Edmund Burke's reaction was also tinged with a certain amount of regret; no more so than after the debate on 25 November when the Rockinghams replied *en masse* to Grenville's speech. 'We went down to the House', Burke noted despondently, 'and in our turn, in the strongest terms, renounced him and all his works. We are since to the great triumph of the Ministry quite loose.'[17]

The poignancy of Burke's words soon became apparent. Convinced that any further attempt to unite the opposition would be futile, Bedford opened negotiations with the ministry on 27 November. Three weeks later his corps entered office. This was not what Grenville had expected and Bedford's decision left him shocked and disgusted. It gave rise to a great deal of acrimonious feeling of which the article in the *Political Register* was but a small part.[18] Grenville remained convinced that Bedford had known of

[14] *Grenville Papers*, iv. 234.

[15] On 26 Nov. Rockingham told Newcastle the report was more or less accurate, B.L. Add. MSS 32987, fo. 120.

[16] *Grenville Papers*, iv. 253.

[17] *Burke Correspondence*, i. 336.

[18] 'A Word of Parting to his Grace the D[uke] of B[edford], Almon, *Political Register*, i. 465–8.

his return to office before the session began, and never forgave him for pretending otherwise. His treachery, Whately explained in April 1768, precluded any further contact between the two parties. Though we may still agree on 'principles', he declared firmly, 'it was of moment to state the utter impossibility of our ever again banding with men whom we could never absolutely trust'.[19] It was not a hollow threat. The broken friendship was never repaired, and debates in the House over the next two years were characterized by an unparalleled bitterness between Grenville and Rigby.[20] Yet Grenville's anger was not due solely to a sense of desertion. This was at the root of his disappointment, but two other considerations magnified the defection beyond its true perspective. The first was the prospect of campaigning alone for the rest of the session. The Bedford party totalled no more than twenty in the Commons, but with Rockingham alienated it was a supplement Grenville badly needed. The debates in January 1767 had shown the pitfalls waging a lone campaign against ministerial policy. The second, and more worrying consideration, was the pending general election. The breach with Bedford could not have come at a worse moment in Grenville's preparation for the poll. He had no electoral influence himself, and depended solely on the help of powerful allies, like Bedford, to ensure the return of his friends. In fact two leading members of Grenville's own party relied for their seats on two of Bedford's followers, Sandwich and Sir Laurence Dundas, and their futures were now in doubt.[21] It is little wonder that Grenville judged Bedford's return to office so harshly.

This gloomy picture was lightened only by the efforts of Grenville's most ardent followers to boost opposition morale. Though they could do little about Bedford's desertion, bridges with the Rockinghams were being built before the year was out. After the argument between Grenville and the Rockinghams on 24 November this had seemed an

[19] *Grenville Papers*, iv. 274.

[20] Grenville's behaviour on this occasion was quite consistent with the treatment of Charles Jenkinson's return to office at the admiralty in Dec. 1766. Grenville 'returned no answer' to the letter carrying the news, and 'Forbid his porter ever to let him into his house again', ibid., p. 393.

[21] Dundas controlled Alexander Wedderburn's seat at Richmond and Sandwich the prospective seat of Henry Seymour at Huntingdon.

unlikely prospect, but Lyttelton and Whately, in particular, were not prepared to let this one outburst ruin the good work of the previous twelve months. The main obstacle to a union between the followers of Rockingham and Grenville was the leadership of the two parties. Since the breakdown of the negotiations in July, relations between Grenville and Rockingham had been coloured by a growing distrust that made co-operation impossible. The root of the animosity had nothing to do with principle. The negotiations of the summer had foundered on Bedford's demand for Conway's dismissal as leader in the Commons, not Grenville's insistence on a firm colonial policy. It was based more on a misconception of the other's motives. Rockingham saw subversion in every move that Grenville made. The guarantee on American policy and places for his friends were not excessive, but in Rockingham's warped view of the negotiation they became annoying restraints that 'neither Lord Temple nor Mr. Grenville had the right to put . . .'[22] These terms were Rockingham's scapegoat after the talks had failed, and long after contact between the parties had ended he still believed that Grenville was the main stumbling-block to a union. No amount of counselling from Hardwicke or Newcastle could soothe this obsession, and in the autumn Rockingham had finally given vent to his frustration by telling the Duke of Bridgewater that 'he would hear of nothing in which there was a Grenville'.[23]

Grenville's view of Rockingham never reached such passionate heights, but his feelings of distrust certainly ran as deep. Grenville had lost faith in Rockingham during the opposition negotiations of the spring and the experiences of July did nothing to restore it. He firmly believed that Rockingham's intention was simply to isolate his party from Bedford and pick off its leading members for the new administration. This ploy Grenville vehemently opposed, for its success would have left him with no power in the Commons and even less over the direction of ministerial policy. Overshadowing this fear there was also a degree of personal disdain for Rockingham's political acumen. Grenville not

[22] This quotation is taken from a memorandum by Dowdeswell, covering the talks of 1767, W.W.M. R1–536.

[23] *Lyttelton Memoirs*, ii. 740. The remark was made at the Newmarket races.

only considered that his ministry had been wrong to repeal the Stamp Act but that Rockingham himself was a poor choice as first lord. The years 1765–6 had proved to Grenville that he had neither the knowledge nor the experience to carry the office successfully.[24] The news of Rockingham's statement to Bridgewater was greeted with little surprise by Grenville. It merely confirmed his preconceived ideas of Rockingham's conduct and the attack of 24 November was Grenville's blunt response.

It was to their credit that Grenville's closest advisers did not become embroiled in this personal dispute. Lyttelton re-opened talks with Burke and Hardwicke for the Rockinghams in late December, and it was clear that despite the intransigence of their leaders much common ground still existed.[25] The immediate problem, as Burke told Lyttelton in their discussion, was the unfinished East India business before parliament. In the second reading of the Dividend Bill before Christmas, which fixed the dividend on East India stock at 10 per cent, the opposition had divided the House but posed no threat at all to the government's position. In June 1767 the ministry had passed legislation in the wake of the East India inquiry restricting the payment of dividends to the 10 per cent sum, as opposed to the 12 per cent favoured by the Company and the opposition. The restriction lasted only for the duration of that parliamentary session, however, and opposition leaders seized the opportunity of renewal in the winter of 1767–8 to do battle with the government again. The grounds for protest would be the same as the spring of 1767: that restricting dividend on stock was an unwarranted interference in the private affairs of a legally constituted chartered company. The debates before Christmas did not go well, with a minority of forty-one votes on 16 December simply underlining the urgent need for unity to prevent a collapse of the opposition campaign.[26] Unfortunately the extent of the discussions between the parties,

[24] There is only one known instance of Grenville voicing a personal attack on Rockingham, it occurs in a letter to Buckinghamshire 23 June 1766, H.M.C. Lothian, p. 261.

[25] See *Grenville Papers*, iv. 240–9 and *Burke Correspondence*, i. 339–40.

[26] The result of the division that took place on 16 Dec. was 128 to 41 votes, *Commons Journals*, xxxi. 492. Background to the Dividend Bill can be found in Sutherland, *East India Company in Eighteenth Century Politics*, chap. VI.

and Grenville's immediate reaction, remains unknown. The first debate on the East India Company after the Christmas recess certainly suggests they were not very thorough. On 22 January the voting on the report for the Dividend Bill was 120 to 25; and two days later, on the third reading, 131 to 41.[27] These low minorities were not the stuff to inspire confidence, but all was not lost. In both debates supporters of Rockingham and Grenville had shared the burden of speaking in opposition. Moreover, Grenville himself had shown no inclination to give up the opposition battle. Attendances and morale were low but it was at this moment of inertia that Grenville's true worth as a party leader shone through. As Winstanley stressed in his study many years ago, no matter how serious the defeat or demoralizing the prospect of opposition, Grenville refused to contemplate a retreat. The remaining weeks of this parliament were to be some of the busiest of his opposition career. Whether Lyttelton's intelligence from the Rockinghams had lessened his anger is not clear, but past conflicts were forgotten as Grenville used every opportunity to attack the ministry on both its imperial and domestic policies. The belief that 'the Grenvilles had no zest' for opposition during this short session could not be further from the truth.[28]

This did not make the quest for union any easier. On certain issues the Rockinghams found it very difficult to support Grenville's line of attack. His endorsement of Beckford's bill outlawing corruption in elections, for example, put Dowdeswell and his colleague in a tricky position.[29] It was not that its main provisions, an oath against bribery and disqualification of non-resident freeholders, called for radical change. The Rockinghams were just as keen to eliminate Treasury influence in elections as Grenville. But the fact was that Rockinghamite support would have been badly affected by any tampering with the existing franchise. To meet Grenville half-way therefore, Dowdeswell proposed on 16 February that the oath and disqualifying clause be replaced by an order forbidding

[27] *Commons Journals*, xxxi. p. 538 and p. 542.
[28] Brooke, *The Chatham Administration*, p. 336.
[29] The bill was introduced in the Commons on 26 Jan. 1768.

'all officers of the revenue' from taking part in elections.[30] Grenville was not enthusiastic at first, but when it was obvious that the bill would not be put to the vote in its original form, fully supported the compromise suggestion. On 19 February the opposition was beaten in committee on the bill, but only by the slim majority of 27 votes.[31] On one topic during these weeks, however, the opposition parties experienced no trouble in presenting a united front. Sir George Savile's motion on 17 February for leave to bring in the Nullum Tempus Bill, produced a degree of unity that had seemed impossible after Grenville's speech of 24 November. The bill itself arose from a grant by the treasury to Sir James Lowther of part of the Duke of Portland's Cumberland property. Lowther had petitioned the Crown, claiming that this land had not been in the original grant to the first duke but simply included without authority by the Portland family. The petition had been upheld because according to the legal maxim *nullum tempus occurrit regi* there was no limit against an action by the Crown. Savile did not dispute the legal basis of the grant. He simply hoped to circumvent it by proposing that sixty years' tenure of property should bar action by the Crown for its recovery.

Plans for the debate were strictly controlled by Rockingham. He was under no doubt of the opposition's strength on this cause and took the utmost care not to reveal his hand. Though messages requesting attendance were sent out, they made no mention of the motion in order to catch the government off-guard. Furthermore, it would appear from the list of speakers that consultations with Grenville and his followers only took place in the Chamber before the debate began.[32] If Rockingham knew that Grenville would support Savile's motion beforehand, he was certainly not going to allow him the limelight he had enjoyed after the land tax victory.[33]

[30] Walpole, *Memoirs of George III*, iii. 112.

[31] The voting figures were 96 to 69 votes. For further information on this see Lawson, Ph.D. thesis, pp. 136–9.

[32] *Burke Correspondence*, i. 345.

[33] Rockingham may have known that Grenville would support the motion before 17 February. In that month's issue of the *Political Register* there is a brief description of Lowther's claim followed by a reference to Grenville's treatment of a similar case when first lord. 'In like manner' runs the report, 'a grant

(cont'd. ...)

It was the type of question that drew the sympathy of the unattached MPs, especially the country gentlemen, and Rockingham fully intended to exploit the opportunity of voicing their disapproval at the government's insensitive action. Much to Rockingham's delight, the debate went according to plan. No one actually challenged the treasury's right to make the grant but the mood of the House was clearly in favour of reforming the law. This put North, recently appointed as leader in the House, in a tight spot. During his speech he 'objected on the impropriety of the time, the very end of a parliament', but was wise enough not to place 'a negative on so popular a bill'.[34] To rid himself of the question North moved instead that the orders of the day be read. This was an unsatisfactory procedure on a well supported point and it obviously angered many MPs present. In a close division the government carried its motion by a mere twenty votes.[35]

A minority of over a hundred provided encouragement for the opposition so close to the end of a parliament. It would not have been possible without the support of the independents, but that does not lessen the value of Rockingham's preparation or the performance of opposition speakers in the debate. The day's events certainly gave Grenville cause for celebration after the dark weeks of November and early December. In the first place the reconciliation with the Rockinghams now seemed complete. Despite the lack of prior consultation, Grenville, and two leading followers, Thomas Pitt and Henry Seymour, had all spoken and voted for Savile's motion without reserve. Second, and more significant, it was clear from the division list that Grenville's party had weathered the storm of Bedford's desertion without suffering severe damage. There are three known sources for the minority of 17 February, and when collated together they paint an optimistic picture of Grenville's party at the end of the parliament. The least satisfactory of the three was

somewhat similar was solicited for when Mr. G[renville] was minister in prejudice of the D[uke] of D[evonshire] but Mr. G[renville] refused to ask for it, saying the attacking of private property in that manner, was a thing too serious'. Almon, *Political Register*, ii. 123.

[34] Walpole, *Memoirs of George III*, iii. 115–16.

[35] The figures were 134 to 114, *Commons Journals*, xxxi. 614.

prepared by Rockingham.[36] It lists only 107 members of the minority under party allegiance, and is fraught with characteristic errors. James Grenville, for example, appears under Grenville's name when it was widely known that he was a devout follower of Chatham. Lord George Sackville, on the other hand, was unclassified when it was equally well known that his sympathies lay with Grenville. The remaining two lists prepared by John Almon and Edmund Burke are much more reliable.[37] Though they make no attempt to ascribe party allegiance, they do provide a sound basis for assessing Grenville's support in the division. Of the names common to both minority lists, twenty-six were Grenvillites. These can roughly be divided into three groups. In the first were supporters whose personal and political ties had bound them to Grenville's cause since 1765: William Mathew Burt, Clive, Bamber Gascoyne, Thomas Grosvenor, William Gerard Hamilton, James Harris, Thomas Howard, Thomas Orby Hunter, Henry Seymour, Henry Shiffner, Francis Vernon, John Walsh, and Thomas Whately. The second consisted of Lord George Sackville and Sir John Irwin who along with John Carnac, elected to the House on 6 February 1768, had joined Grenville's party after 1765.[38] The third was made up of independents whose loyalty had remained firm despite Bedford's decision to join the ministry: Richard Wilbraham Bootle, William Throckmorton Bromley, Sir Robert Burdett, Assheton Curzon, Sir Charles Hardy, Anthony James Keck, Sir Roger Newdigate, Matthew Ridley, John Rushout, and Jarrit Smith.

A voting strength in the mid-twenties was not outwardly impressive, but there is no doubt that Grenville breathed a sigh of relief at the number. Since July 1767 his parliamentary fortunes had suffered a severe reverse. Relations with Rockingham had reached their lowest ebb and his closest allies, the Bedfords, were now in office. Worse still, Grenville's chances of an early return to office appeared to have evaporated. The vote on *nullum tempus* was the first real sign that this depressing series of events had come to

[36] W.W.M. R81–218.

[37] Almon's list of the minority appears in *Debates*, vii. 365–8, and Burke's amongst his papers at Sheffield, Burke MSS 11/8.

[38] Clive MSS vol. 56.

an end. Even though Grenville's party had seen a reduction over the past three years, to take almost 25 per cent of the minority vote so late in the session was no mean achievement.[39] He not only had the usual antipathy towards end of session debates to contend with but also the anxiety amongst his supporters over the coming election. Many were in difficult seats that demanded a rigorous canvass, and they simply could not afford to spend the weeks after the Christmas recess in London. The fact that many did stay and vote with Grenville was a tribute to his powers of leadership, and the result could not have provided a greater boost to morale in the elections for the new Parliament.

Parliament was dissolved on 11 March, and Grenville set off for the country in need of all possible encouragement. The general election would provide the severest test of his leadership to date, and the prospect offered little cheer. Without direct influence over the electoral fortunes of his followers, Grenville was forced to spend the month or so of polling in frustrating idleness at Wotton. At a time when he should have been most active, there was nothing he could do but wait patiently for information of the returns. This does not mean that Grenville played no part at all in the election. Though polling officially began in March, he had been canvassing on behalf of his friends long before this date. It was a limited and unrewarding role based solely on his personal influence with political allies at Westminster. On very rare occasions his pleas were heeded. In January 1767, for example, he asked Lord Aylesford to use his interest at Guildford to support the candidature of Sir Fletcher Norton. 'He assured me of his readiness to comply with my request and to give you any assistance in his power at Guildford', Grenville told Norton after the interview, 'if you can point out any way in which he or Lord Egremont's family can be of use to you. I am persuaded they will make no difficulty.'[40] His 'persuasion' was proved correct, for nothing untoward did occur before the election. Norton's own interest in the borough allied with that of the Aylesford and Egremont families ensured an unopposed return in 1768.

[39] This revises the figures presented by Brooke, *The Chatham Administration*, p. 247.

[40] Grenville Letterbook, vol. 2.

This minor success, however, was not typical of the election as a whole. More often than not Grenville's interference was the precursor to a lost cause. Two cases in particular are worthy of note. The first was the failure of Robert Mackintosh to secure a seat in the Scottish constituency of Perth Burghs. Mackintosh was an associate of Clive and through him became known to Grenville. His interest in the seat appears to have begun in 1766 when he purchased a large estate in the vicinity of Perth, but cultivation of the five burghs in the constituency never ran smoothly.[41] From the outset he was strongly opposed by the sitting MP, George Dempster, and in a bitter struggle for ascendancy over the electorate both candidates readily sought help from more powerful allies. Grenville's contribution to Mackintosh's campaign was prompted by a request from Clive to lobby two politicians with an interest in the area. 'If Sir William Duncan or Sir Laurence Dundas have any interest in the Royal Burghs in question', he replied to the plea for help, 'I will apply to them to give their assistance to Mr. Mackintosh.'[42] Grenville dutifully carried out the promise, writing at least four letters on his behalf. But it had little effect on the outcome. Though Dempster was forced to withdraw his candidature shortly before the poll, his replacement, William Pulteney, won a comfortable victory. The second concerned the inability of Sir Hew Dalrymple to find a seat for election to the new parliament. Dalrymple was a personal friend of Grenville, and turned to him in the autumn of 1766 when he realized that his return for the next parliament at Haddingtonshire might be in jeopardy. Though his family did have considerable influence over the constituency, a rival interest under the Dundas connection threatened to block his adoption. Dalrymple's immediate reaction to this set-back had been an attempted bargain with the ministry. But his offer of political support in return for help in the election was coldly received. He then sounded Grenville and found, much to his delight, that he was more than willing to help. 'I will take the first opportunity', he told Dalrymple on 24 October 1766, 'of

[41] Zetland MSS ZNK X 1/2/90.
[42] Grenville to Clive, 4 Sept. 1767, Grenville Letterbook, vol. 2.

applying to my brother Lord Temple for his good offices with Captain Christie at the ensuing election and hope by that means that his inclinations to assist you will be determined in your favour.'[43] This was sufficient to soothe Dalrymple's anxiety for the next twelve months but by the autumn of 1767 his position was so weak that he again asked Grenville for help. In a last desperate throw, Grenville responded by appealing to the sitting member, Andrew Fletcher, to either stand again or retire in Dalrymple's favour.[44] It was not a pleasant task. Fletcher also supported Grenville in parliament, and if either course of action was carried through the party would be sure to lose one member. In the end it lost both. Fletcher had tired of constant disputes over the seat, and refused to stand or endorse Dalrymple's candidature. Unable to exert any further influence, Grenville could only look on as the Dundas candidate was elected.

Such failures were certainly depressing, but Grenville could at least draw solace from having played a small part in their election campaign. The majority of his followers lost in the election were obliged to fight their battles alone. In part this was purely design. Seven independent supporters of the party declined the poll, and none consulted Grenville before making their decision: Sir Walter Wagstaffe Bagot, Sir Robert Burdett, Sir James Dashwood, Sir Thomas Pym Hales, Sir Charles Hardy, Jacob Houblon, and Jarrit Smith. There was no common factor to explain their abdication. Some, like Smith, had simply grown weary of politics at Westminster and made no attempt to re-enter the House. Others, like Burdett, found themselves in such a weak position before the poll that they were forced to stand down. Whatever the motive, Grenville was quite helpless to prevent their loss. He wielded no personal or political influence over these MPs, and remained ignorant of the reasons behind their decision not to stand. The cause of their reticence was a mystery, he told Fife after the election, but he suspected that it stemmed from 'having a very melancholy prospect of public business before them'.[45]

[43] Buccleuch MSS GD 110/940/38.
[44] Grenville Letterbook, vol. 2 (16 Feb. 1768).
[45] Ibid., (28 May 1768).

Four other Grenvillites were not afforded the luxury of such a choice: William Mathew Burt, Henry Shiffner, Francis Vernon, and Sir Armine Wodehouse. They represented difficult constituencies and lost their seats in contested elections. Of the four, Burt, Shiffner, and Wodehouse all made appeals to Grenville for help before the election which he was unable to meet. In the case of Burt and Shiffner this was not due to neglect. Grenville simply did not have any influence with the patrons of the boroughs for which they sat. Wodehouse's defeat in Norfolk was more complicated. His trouble began in the summer of 1767 when a newspaper in Norwich began attacking his record in parliament over the previous five years. He told Grenville of the campaign against him, and in reply received an immediate declaration of support. It was impossible to believe, Grenville wrote, that he should now be running into trouble for his votes in favour of general warrants, the Stamp Act, or even the reduction of the land tax. The attacks were nothing but 'infamous vehicles of falsehood', he observed dramatically, 'and I hope there is no friend of mine in that country who will not give to you the strongest testimony of his approbation'. Grenville himself was under no doubt about Wodehouse's worth, and wrote to all his influential contacts in the country 'to assure them of the interest I take in your success'.[46] On one front, however, Grenville made no progress. His most powerful and influential supporter in Norfolk, the Earl of Buckinghamshire, abandoned Wodehouse and endorsed the candidature of two of his opponents, Wenman Coke and Sir Edward Astley. Both were Rockinghams, and there is no logical explanation for Buckinghamshire's behaviour in party terms. In all other contests where he held an interest, Buckinghamshire gave unstinting support to the Grenvillite candidates. It was, as Brooke comments, merely direct evidence of the fact that 'local allegiances counted for more than national politics'.[47] There were many examples of this in the election as a whole and Grenville's followers were guilty of returning at least two other adversaries. The first was at Maidstone where the Rockinghamite, Robert

[46] Ibid., (10 Oct. 1767).
[47] *House of Commons*, i. 339.

Gregory, was elected on the interest of Grenville's friend Lord Aylesford.[48] The second occurred at Old Sarum where the Butite John Crauford was returned on Thomas Pitt's recommendation.[49] Local commitments notwithstanding, Grenville was shocked and hurt by Buckinghamshire's action. Wodehouse had held Norfolk in five parliaments and his defeat by Sir Edward Astley and the fourth candidate, Thomas De Grey, prompted Grenville to deliver a stern lecture to Buckinghamshire on his responsibilities to the party. 'I told you with great truth', he wrote tersely, 'that I was very sorry Sir A. Wodehouse had lost the election for the county of Norfolk and when you consider the constant regard and friendship which that gentleman has expressed towards me and the agreement of his sentiments with mine upon all public business you cannot I think wonder at it.' Grenville was particularly angry at the way his instructions had been ignored. 'As to the conduct of my friends upon this occasion', he chided Buckinghamshire, 'I did desire them to give their best support to Sir A. Wodehouse.' This was not done hastily or without prior consideration. 'I will freely own to you', Grenville added, 'that if I had a vote in that county myself I should never have given it in favour of any one who had founded his interest and hopes of success upon so notorious a misrepresentation as that which was contained in Sir E. Astley's and Mr. Coke's public advertisement concerning General Warrants.' It was a regrettable incident, but fortunately it caused no ill feeling. 'Though your ideas have been different from mine upon the subject of this election', Grenville wrote in conclusion, 'I can with the greatest truth assure you that I never conceived any unkind thought of you upon that account.'[50]

The remainder of Grenville's followers lost at the election chose not to stand. There were eight in this group and their reasons for declining the poll differed considerably. One, John Dickson, did not live to see the dissolution. He died in December 1767 and his seat then fell into the hands

[48] The election of Gregory must have caused annoyance, for on 15 Dec. 1767 Grenville vainly recommended Fordyce, a Scottish banker, for the seat, assuring Aylesford that his public opinions agree with ours', Grenville Letterbook, vol. 2.

[49] *Caldwell Papers*, 11. (ii), 135-7.

[50] Grenville Letterbook, vol. 2 (16 May 1768).

of a government supporter. A second Scottish MP, James Abercromby, was forced to stand down because his constituency at Clackmannanshire shared alternate representation at Westminster with Kinross. Grenville did make tentative enquiries about another seat for him in 1765, but they appear to have borne no fruit.[51] Four of the group retired after an unsuccessful canvass: James Edward Colleton, Sir Hew Dalrymple, Bamber Gascoyne, and Robert Knight. All four were known to Grenville personally and their loss was a severe blow to the parliamentary campaign. Indeed, as party leader Grenville was certainly guilty of neglect in allowing such diligent supporters to fall by the wayside. He knew that Colleton and Gascoyne represented government seats and would stand no chance of re-election. Through his friendship with Knight he must also have been aware that the constituency of Great Grimsby had chosen a different candidate. On no occasion during preparation for the election, however, did Grenville interfere on their account or seek alternative seats. Whether or not he was simply waiting for an invitation to act on their behalf is unknown, but there is no doubt that the experience soured Grenville's relations with all three. In 1770 Gascoyne and Knight were returned to the House, but they had abandoned the opposition for the security of safe government seats. Two years later Colleton followed suit when he was elected for St Mawes. The final two members, Andrew Fletcher and John Sargent, did not campaign for re-election at all. Fletcher retired to his country estate at Saltoun, and Sargent concentrated on his banking interests in the City. Both had played little part in the 1761 parliament, and any sense of loss felt by Grenville was purely on a personal level.

In fact there was no need to brood over any of these casualties. Nineteen supporters had been lost at the polls, but Grenville knew well that it could have been many more. At the turn of the year the prospects of fighting a successful election campaign had looked bleak. Of the seventy or so MPs carried into opposition in 1765, several had drifted away from Grenville, and eleven were actively supporting the government. Moreover, his party had been deprived of

[51] Namier and Brooke, *House of Commons*, iii. 404.

Bedford's crucial electoral influence at a most difficult period in its relations with the Rockinghams. It is not surprising therefore that in his assessment of the election Grenville could barely disguise his relief at the results. Despite some notable exceptions, he told Fife in May, 'my friends in general have succeeded in their elections where they have stood'.[52] It was no idle boast; the hard core of the party had survived election. But there was little in this success for which Grenville could take the credit. An examination of Grenvillites returned to the new parliament reveals that they still relied for their seats on their own resources or those of Grenville's powerful allies. In the former category there were nine members: James Duff (Earl of Fife), William Gerard Hamilton, James Harris, Thomas Orby Hunter, William Hussey, Sir Fletcher Norton, Percy Wyndham O'Brien (Lord Thomond), Thomas Pitt, and James Wemyss. In the latter there were eleven, with Clive controlling by far the largest bloc. Apart from his own return at Shrewsbury, five of his associates were re-elected to the House: Henry Crabb Boulton, John Carnac, George Clive, Richard Clive, and John Walsh. Further contributions were made by Suffolk who ensured the safe return of Thomas Howard and Thomas Whately; Lord George Sackville who arranged his own re-election and that of Sir John Irwin; and Buckinghamshire who secured the seat at Bere Alston for George Hobart. In addition there were two returns that owed nothing to the influence of Grenville or his patrons. The first was Alexander Wedderburn's re-election at Richmond on the interest of Sir Laurence Dundas. Some doubt had been cast over his return after December 1767, for despite a personal attachment to Grenville, Dundas's political connection lay with the Bedfords. Wedderburn had sensed the danger of losing his support in September 1767, and pleaded with Grenville to pay him a visit.[53] No record of a personal call exists, but it is fair to assume that Dundas was not offended by Grenville's conduct in the winter of 1767–8. Wedderburn's candidature was accepted and he continued to represent Richmond until the dispute over

[52] Grenville Letterbook, vol. 2 (28 May 1768).

[53] The message was sent through Whately, *Grenville Papers*, iv. 160–2.

Wilkes in the spring of 1769. The second was Henry Seymour's return at Huntingdon. He had represented Totnes in the last parliament, and owed this new seat to the interest of Lord Sandwich, his half-brother. Sandwich was a leading Bedfordite, and there is no doubt that he could have made life very difficult for Seymour after the split with Grenville in December 1767. Fortunately, a payment of £800 and strong family ties overcame the political differences, and Seymour continued to sit for Huntingdon until 1774.

The return of so many close friends was a commendable achievement for a party with limited electoral muscle. Even in his private thoughts it is unlikely that Grenville had envisaged a better result. He had certainly had his share of good fortune at the polls and it did not end with Seymour's return at Huntingdon. As well as the twenty-two MPs re-elected from the 1761 parliament, his party gained nine new recruits. Of these only one, Sir Robert Fletcher (MP for Cricklade), was responsible for his own return. His attachment to Grenville is interesting because at the time of the election he was *persona non grata* with Clive. He had been cashiered from the East India Company's service after an abortive mutiny over the double *batta* in 1766, and despite several subsequent requests from Fletcher, Clive refused to reopen the case. In 1769 Fletcher finally appealed to Grenville to intercede in the dispute, and was rewarded in the summer with full restoration to his former rank in the Company. There appear to be three main reasons behind Grenville's action on Fletcher's behalf. First, he was a personal friend of some long standing. Grenville had recommended him to the Company's service in 1764, and felt partly responsible for Fletcher's misdemeanour.[54] Second, Grenville believed that Fletcher's considerable military and administrative talents should not be wasted for the sake of a personal vendetta. 'I should not have applied to you at all concerning Sir Robert Fletcher', Grenville told Clive on 28 May 1769, 'if I had not thought it consistent with and subservient to your own honour and character.'[55] Third, and most important of all, Grenville did not want

[54] S.T.G. Box 21 (62).
[55] Powis MSS G37, Box 57.

to lose Fletcher's support at Westminster. He valued their friendship and sensibly seized the opportunity to draw Fletcher into his political circle. By July 1769 the bond between them was firm. 'Should I embark for India next winter', Fletcher told Grenville in confidence, 'I think I can easily place Mr. Wedderburn or any other person in Cricklade.'[56]

The allegiance of the remaining newcomers is much easier to discern. Grenville's younger brother, Henry, came in for Buckingham on the interest of his elder brother, Temple. Henry Grenville was a diplomat of some note and had resigned his previous seat at Thirsk while abroad. Clive ensured the return of two more close friends: William Clive, his brother, and Henry Strachey, his personal secretary. William was elected on Clive's own interest at Bishop's Castle and Strachey on that of John Walsh at Pontefract. Three further returns were secured by Grenville's friends in the Lords. Buckinghamshire brought in Thomas Durrant for St Ives; Suffolk returned Arthur Chichester (Lord Donegall) for Malmesbury and Lord Lyttelton arranged the election of his son, Thomas, for Bewdley. Of all the new arrivals, however, Thomas Hampden (MP for Lewes) and Thomas Fenwick (MP for Westmorland) were the most fascinating. In rare examples of opposition co-operation, their returns arose from a combined effort by supporters of both Rockingham and Grenville.

Hampden's election provided most comfort for Grenville because he was the son of an old friend and admirer, Baron Trevor. Though Newcastle held nominal control of Lewes, Trevor had arranged his son's nomination in August 1767 through his brother, the Bishop of Durham. He owned considerable property in the borough and, in contrast to the rest of the family, was also a good friend of Newcastle. It proved to be a timely intervention on Trevor's part. The duke's influence over the constituency had waned in the 1761 parliament and he was more than willing to accept Trevor's support and nominee. Nevertheless the suddenness of this favourable response did take both Grenville and Trevor by surprise. On 2 August Trevor told Grenville that

[56] S.T.G. Box 21 (63).

despite the lack of competition, it would be better to wait until Hampden had toured the constituency before predicting the outcome. Grenville was in full agreement. 'I rejoice very sincerely at the account which you give me of the arrangements taken to bring Mr. Hampden into Parliament', he replied guardedly, 'the Duke of Newcastle's consent and concurrence will I hope preclude any opposition being given to a nomination so proper.'[57] In the end their anxiety proved ill-founded. Newcastle fulfilled his promise to endorse Hampden's nomination and his canvass was an unreserved success. 'My son has not met with a single negative', Trevor reported on 24 August 1767, 'no competitor has yet been, nor, I hope, will be started.'[58] There was only one further scare: in the winter of 1767–8 Newcastle fell ill, and Trevor hurried down to Lewes in case of trouble.[59] It was a wasted journey. Hampden encountered no unforeseen opposition to his candidature and was returned at the head of the poll with a majority of seventeen votes over the losing candidate, Thomas Miller.

Fenwick's election for Westmorland was more complicated, and involved both local and national issues. Grenvillite interest in the seat was the responsibility of the Earl of Suffolk who owned two large estates in the county. He had put forward Fenwick's name for the seat in June 1767, and secured his nomination after arranging a deal with Rockingham's friend, the Duke of Portland.[60] The key to the agreement was a shared hostility to the influence of Sir James Lowther, Bute's son-in-law. Lowther was the largest landowner in the two counties of Cumberland and Westmorland, and despite being unpopular with his tenants, had set his heart on returning ten MPs to the new parliament. The opposition to this was simple and straightforward. In return for a promise of the Egremont interest in Cumberland, Portland agreed to back Suffolk's nominee against Lowther's candidates, Robinson and Upton, in Westmorland. It was a fair settlement for both parties. The Egremont interest,

[57] Grenville Letterbook, vol. 2.
[58] S.T.G. Box 22 (40).
[59] *Grenville Papers*, iv. 205.
[60] There are two letters from Suffolk to Portland relating to this deal on 26 and 27 June 1767 in the Portland MSS, PwF 5521 and PwF 5522.

handled by Lord Thomond the young earl's guardian, was not extensive, but it could have determined the result in a straight fight between Lowther and Portland. The duke's role in the Westmorland election, on the other hand, had little to do with his power over the electorate. The family holdings in the country were negligible in comparison to those of Suffolk and Lowther. His strength lay in the fact that he was a respected figure in both Cumberland and Westmorland, and endorsement of Fenwick's candidature would be an invaluable asset at the hustings.[61] The most satisfying aspect of the election for Grenville, however, was the evidence of popular support for the oppositions' campaign at Westminster. Apart from Portland's backing, Fenwick's return had also rested on the crucial support of the influential Wilson family. This interest was in doubt until the last minute when Isaac Wilson, the family spokesman, explained to Upton that he would be supporting his opponent. His reasons for this were a mixture of local and national prejudices that would have elicited Grenville's full approval. In the first place there was Sir James Lowther's overbearing behaviour with his constituents. At the last election, Wilson told Upton, he made a point 'of bringing all the members into the House in his power at a prodigious expense in order to carry any point adopted by the ministry though never so much disapproved by the generality'. The issue he had in mind was the debate on *nullum tempus*. Lowther's opposition to Savile's motion, he continued, carried 'the apprehension of many an aim at being more arbitrary than is consistent; an instance whose of is the affair of Inglewood which gave me great concern when I first heard of it'. Second, he strongly objected to Lowther's connection with Bute. It was Bute, in Wilson's eyes, who was behind Lowther's intemperate actions, and he told Upton not to expect his support until the alliance was broken. 'I have been very sensible for some time', he observed coldly, 'that the connections and measures of him under whose influence you and John Robinson must act were become so very disagreeable to the generality of people

[61] Bonsall, *Sir James Lowther and Cumberland and Westmorland Elections 1754–1775*, chap. xii.

especially in our district, that they would never remain easy whilst Sir James Lowther had two members of his own Choice.'[62] Whether Suffolk knew of Wilson's personal views or not, there is no doubt that he exploited the strong feeling against Lowther in the campaign literature. Fenwick's handbills and circulars appealing to the 'independent freeholders' secured his election by eighty-one votes over Upton.

Nine gains out of a total of 167 new MPs was an insignificant sum but Grenville did not despair at the result of the poll. Though he had lost some close friends in the election, he had a group of thirty-one MPs in the new parliament that included all the party's leading figures. Indeed there was a marked continuity in the structure and appearance of his following from the last parliament. The only notable change was the increasing influence of Grenville's patrons over membership of the party. Of the thirty-one members, Buckinghamshire, Clive, Sackville, Suffolk, and Temple were responsible for the return of seventeen. Within this subdivision Clive's associates again constituted the largest bloc. In fact it could be said that Shropshire and the border counties became something of a power base for Grenville in the years 1768–70. Only Wiltshire which returned five members of his party in the election offers serious competition to this claim. The unknown quantity in the new parliament was how the independent members would react to Grenville's continued presence on the opposition benches. Twenty-eight elected to the 1768 parliament had sat before the dissolution and voted with Grenville in opposition. Yet of these it would seem that Grenville could count on the support of only three: Thomas Grosvenor, Edward Kynaston, and Richard Lowndes. His confidence in Grosvenor and Kynaston was the result of a personal friendship; that in Lowndes stemmed from a neighbourly respect. Lowndes represented the county of Buckinghamshire, and to judge from a letter to Halifax on 21 April 1768, he and Grenville were in accord on political matters. As to Mr Lowndes, Grenville wrote, 'I have lived upon a footing of civility and goodwill and we have always acted together in public business.'[63]

[62] The letter is undated but was probably written on the eve of polling in early April. It can be found in Fenwick's electoral papers in the Upton MSS Box 4, 30.

[63] Grenville Letterbook, vol. 2.

Beyond this Grenville would not commit himself. The votes of the remaining twenty-five unattached MPs would depend, as it had in the last parliament, on the issues before the House.

Fortunately Grenville did not have long to wait for a topic upon which to unite his followers with the independent MPs. In the spring of 1768 John Wilkes secured his election for Middlesex, setting off a train of events that were to dominate opposition politics for the next two years. The problems arising from his return were twofold. The first, and most simple, was to deal with the civil disorders sparked off by his success at the poll and subsequent imprisonment for crimes committed in 1763–4. The second, and more profound, concerned the legal and constitutional implications of Wilke's election to the House while he was still an outlaw and under sentence. In the short spring session of the new parliament, it was the former that occupied Grenville's thoughts. The rioting in London had certainly taken the government by surprise, and Grenville immediately laid plans to exploit ministers' mishandling of the situation. It was done in great secrecy with Suffolk and Whately taking a leading part in directing strategy.[64] The attack began in the Lords on 12 May when Suffolk attempted to embarrass ministers by moving 'to address the King to confer some mark of distinction on the Lord Mayor Harley for his activity and spirit during the disturbances'. It was a lightly veiled censure of the government's unpreparedness, and Grenville took up the point the following day. In determined mood he 'painted the supiness of the ministers', as Walpole put it, 'in strong colours'.[65] On a tactical level he could not have chosen a better subject on which to open the opposition campaign. It not only attracted the sympathy of the unattached MPs but also drew active support from the Rockinghams. 'People are sleeping in government', Dowdeswell declared, echoing Grenville's sentiments, 'yesterday and today tumults have increased.'[66] Fear of lawlessness in the streets of London ran deep on all sides of the House, and Harley's strong measures were unanimously approved.

[64] *Grenville Papers*, iv. 283–6.
[65] Walpole, *Memoirs of George III*, iii. 142–3.
[66] B.L. Egerton MSS 215, p. 21.

Without a division it is difficult to measure the true impact of Grenville's attack. It is possible, as Walpole infers in his account of these proceedings, that the government would have taken the initiative on Harley and the continued rioting without prompting from the opposition.[67] Nevertheless this energetic opening to the new parliament left no one in doubt that Grenville intended to pursue the opposition campaign with vigour. The letters exchanged between Grenville and his followers during the summer months displayed a confidence that had not been evident since the debates on the East India Company of the previous year. The Rockinghams were certainly impressed with Grenville's role in the debates during this short session, and reopened contact with representatives from his party in early July. Though it was the followers rather than the leaders of the two groups who bore the main burden of negotiation, the talks went well. In the initial round Wedderburn and Burke discussed the idea of formal co-operation and found enough common ground to be optimistic of the future. Indeed, Wedderburn was moved to tell Whately after the conversation that 'he saw a disposition' in the Rockinghams to follow Grenville's 'lead in the House of Commons'.[68] This was a gross exaggeration, but Burke's new-found enthusiasm for the Grenvillites does bear examination because it gives a unique insight on his attitude to the party. At root it was Grenville's handling of two recent political crises that had impressed him most of all. The first concerned Sir Fletcher Norton's desertion to the ministry shortly after the election. This had arisen from an offer to supervise the government's legal affairs in the Wilkes case. At first Norton had been undecided, but in mid-April he chose to ignore the wishes of his friends and threw in his lot with Grafton. The loss of his debating talents was a severe blow, and Burke had only praise for Grenville's conduct during the unhappy incident. 'Mr. Burke took notice that the language which he heard you hold, was that of a very wise man', Whately reported, 'no Minister could be safe or be active, who was not sure of the King, and of the persons with whom he was connected,

[67] *Memoirs of George III*, iii. 143–5.
[68] *Grenville Papers*, iv. 311. The quotations that follow are taken from here.

which he had been told had been a principle you had such insisted on lately.' The second related to the unexpected re-entry to office of the Bedfords in December 1767. It had been a stiff test of principle, and Grenville's stern response to their treachery demonstrated to Burke that he was 'certainly a most excellent party-man'. In his analysis of December's events Burke could find no blemish on Grenville's record. 'Your behaviour to the Bedfords had proved you might be relied on', wrote Whately, 'that you would not desert those who would abide by you, and were steady to all your purposes; that it was pleasant to be connected with such a man, and the party would act with confidence who acted under him.'[69]

Such statements certainly boded well for the coming session, but they did not constitute the basis of an opposition union. Though Burke's pronouncements promised a great deal, he was not the leader of a party and Grenville sensibly viewed them with caution. Wedderburn saw Rockingham in the autumn and he was far more reserved in his commitment to co-operation. On colonial affairs he was willing to forget all previous disagreements. 'His Lordship's language', Whately wrote in his account of the meeting, 'was that it was impossible the several parts of the Opposition could differ now on American measures; that in Opposition they should entirely agree.' The question of office, however, was quite another matter. If the two parties were ever 'called upon to form a Ministry', Rockingham declared, 'as Mr. Grenville would naturally expect the first place, there would be difficulties in giving securities to his Lordship's friends'.[70] Whether Burke liked it or not Rockingham was not ready for a formal alliance with Grenville. He had no objection to a pragmatic arrangement on the floor of the House but did not wish to concert a plan of attack before the session began. Grenville expressed no regret at this attitude. He was in confident mood after the debates of May, and it was the government not the opposition who had most cause for concern about

[69] These criteria for party unity appeared almost unchanged in Burke's pamphlet, *Thoughts on the Present Discontents*, published in 1770. See, for example, the passages on the necessity of loyalty in the ranks in Pollard, ed., *Political Pamphlets*, pp. 333–4.

[70] *Grenville Papers*, iv. 392 (28 Oct. 1768).

parliamentary policy. In October both Chatham and Shelburne had resigned from the ministry, making it imperative that Grafton find new blood. Grenville's own following had been approached in early November but with little effect. 'The Bedford party are industriously endeavouring to detach his friends from him', Mrs Grenville noted in her diary on 26 November, but only the Earl of Powis 'had been gained by these means'.[71] Furthermore, Chatham's resignation had been the signal for an advantageous family reunion. Temple visited his brother-in-law at Hayes in November and early December, and, with Grenville's blessing, arranged an amicable settlement of their past differences. It was an ingenious move Though there were no binding political commitments on either side, Chatham's small but able following in the Commons had now been secured for the opposition. The core of the old Cobham group was reunited after seven years of acrimony and misunderstandings.

Grenville's self-assurance was only partly borne out by proceedings at Westminster during December and January. The opposition had the popular issue of Wilkes upon which to build its attacks, but failed initially to capitalize on the government's indecision in executing its policy. Grenville himself must take a share of the blame for this. Before it became absolutely clear that the government intended to deprive Wilkes of his seat at Middlesex Grenville's attitude to his election was coloured by two overriding considerations. First, would his approach to the problem appear consistent with the actions he had taken over general warrants in 1763–4? Second, could his opposition to ministerial policy be reconciled with a strict legal interpretation of all the issues surrounding Wilkes's election? In practice these questions were to cause his followers no little confusion. On 19 December, for example, the Commons discussed a libellous article written by Wilkes and published in *The St. James's Chronicle* on 10 December 1768. It was merely a preliminary to Wilkes's expulsion, but Grenville, supported by Sir George Savile for the Rockinghams, took a strong line against the government's plans to proceed without the chief witness. They argued most forcibly that if the

[71] Ibid., p. 404.

government insisted that the House rather than the courts pass judgement on the article then Wilkes would have to be present. Support for their objections is unknown because the challenge was defeated without a division.[72] Nevertheless Grenville's followers could have been forgiven for believing that here was an open declaration of war against ministerial policy on Wilkes's election. His speech was unequivocal and he continued the same line of attack on 19 December when other witnesses in the libel case were examined. Such assumptions, however, soon proved ill-founded. On 27 January a petition from Wilkes concerning abuses in his arrest of 1763 came before the Commons, and in this debate Grenville supported the government. The point at issue was only a technical motion by the ministry restricting discussion of the petition, but Grenville endorsed the proposal for, in his words, 'this complaint ought in point of law and privilege to be confirmed'. All other questions arising from Wilkes's arrest in 1763 had been clarified, he told MPs bluntly, and desired to know 'an instance where a subsequent House of Commons have taken up a breach of privilege, and that the parties should come to be tried a second time'.[73]

Grenville saw no contradiction in his conduct during these debates. He had made a close study of the legal and constitutional questions involved in the case and acted accordingly. At no time during the battles over the Middlesex election did he let Wilkes's personal campaign against the ministry cloud his judgement on the issues raised by his return. To Grenville this was an irrefutably logical standpoint, but the distinction did escape many of his followers. In fact on the motion of 27 January to restrict discussion of Wilkes's petition Grenville was dismayed to find himself at odds with members of his own party. In an embarrassing show of disunity, several of his followers supported the government while an equal number voted in the minority. There seems little to say in Grenville's defence for this confusion. An examination of the minority vote merely points to the fact that he had not communicated his views to every member of the party.[74] The opposition votes of

[72] B.L. Egerton MSS 216, pp. 46-71.
[73] Ibid., pp. 177-217.
[74] *Gazetteer*, 1 Feb. 1769.

Thomas Fenwick and Thomas Hampden could perhaps be excused on the grounds that they had only come into the House at the last election. There can, however, be no acceptable explanation for the conduct of seven others: Henry Crabb Boulton, Henry Grenville, William Gerard Hamilton, George Hobart, Thomas Howard, William Hussey, and Henry Seymour. They were all known to Grenville personally and should have been aware of tactics during the debate. Seymour's vote represented a major blunder. He was a member of Grenville's inner circle and played an important rôle in evolving party strategy at Westminster. Grenville's independent supporters did not suffer the same agonies. Only seven who had supported his cause in the last parliament cast their vote in the minority: Sir Thomas Bunbury, Edward Herbert, Anthony James Keck, Richard Lowndes, Matthew Ridley, John Rushout, and Robert Waller. Of these it is likely that Lowndes alone would have known of Grenville s intentions prior to the debate. The rest simply decided to vote on the merits of the arguments and suffer the consequences which were never in doubt. The government carried its motion by the comfortable margin of 278 to 131 votes.[75]

Attempts to eradicate these internal divisions were not entirely successful. Wilkes's libel next came before the House on 2 February, and Grenville again discovered his friends on opposite sides in the division. On this occasion, however, the damage was superficial. Grenville's opposition to the government's policy on the article in *The St. James's Chronicle* had been established before Christmas and he was able to carry almost two-thirds of his party into the minority. These included all those who had voted in opposition on 27 January with the addition of: John Carnac, Arthur Chichester, Clive, William Clive, Thomas Durrant, Fife, Sir John Irwin, Henry Strachey, and Thomas Whately. Of those missing from the minority only Thomas Pitt is known for certain to have voted with the government. It is possible nevertheless that Wedderburn also supported the administration in deference to his patron Sir Laurence Dundas. Though not wholly satisfactory, the division

[75] *Commons Journals*, xxxii. 156.

represented a creditable recovery from the débâcle on 27 January. At this point Grenville commanded no more than thirty votes in the Commons and the majority of these had fallen into line after his speech. If there was one disturbing aspect it concerned the failure of the independents to support his attack. Only six unattached MPs, previously sympathetic to his opposition campaign, voted in the minority: Sir Brook Bridges, Sir Walter Blackett, Edward Herbert, Anthony James Keck, Matthew Ridley, and John Rushout. No pattern had yet emerged, but it was becoming clear that one section of his supporters was not impressed by his strict legal interpretation of the Wilkes issue.

This fact was painfully confirmed the following day when Barrington moved for Wilkes's expulsion. In a debate full of fine oratory, Grenville severed the bond with the majority of his independent supporters by strongly opposing the motion. There was no sentiment expressed as the last rites were given. The arguments were based firmly on legal precedent and took little note of MPs' personal animosity towards Wilkes. If Grenville was anxious to preserve independent support, it would not be at the expense of his consitutional principles. In the division the depressing trend towards supporting the government continued unabated.[76] Seven long-standing admirers in all abandoned his opposition campaign on this day: William Drake, Sir John Glynne, Edward Kynaston, Thomas More Molyneaux, Sir Roger Newdigate, Robert Waller, and Sir Charles Tynte. None returned to the fold before Grenville's death in November 1770. There are two other division lists on the Middlesex election for this session and they both reveal a gradual destruction of Grenville's independent support.[77] The defections were of two types; MPs who had transferred their allegiance from opposition to government and those simply declining to vote at all. In the first category there were eleven unattached MPs: Sr. William Bagot, Sir Thomas Bunbury, William Drake, Sir John Glynne, Edward Kynaston, Richard Lowndes, Thomas More Molyneaux, Sir Roger

[76] The following analysis is based on the two minority lists in *Gentleman's Magazine*, 1769 pp. 631–7 and *The North Briton*, 15 Apr. 1769.

[77] These refer to 15 Apr. (*London Magazine*, 1769, pp. 270–1) and 8 May (*The North Briton*, 27 May 1769).

Newdigate, Sir John Turner, Robert Waller, and Sir Charles Tynte. Their loss was a severe personal and political blow to Grenville. Kynaston and Newdigate were close friends and the others had all been active supporters of Grenville in the House. In the second category there were seven MPs: Sir Richard Warwick Bampfylde, Richard Wilbraham Bootle, William Throckmorton Bromley, Assheton Curzon, Charles Gray, Sir Edmund Isham, and Sir Charles Mordaunt. Of these only Bampfylde was known to Grenville personally but this did not compensate for lost votes. All seven had voted with the opposition in the 1761 parliament and Grenville's party could ill afford their indifference.

It was a gloomy picture offering little hope of improvement. Grenville did retain the support of nine unattached MPs from the old parliament, but only two, Thomas Grosvenor and John Rushout, had had any previous contact with Grenville. The remaining seven were independent MPs over whom Grenville had no hold whatsoever: Sir Walter Blackett, Sir Brook Bridges, Anthony James Keck, Matthew Ridley, William Sharpe, John Rolle Walter, and Thomas Willoughby. This is not to say that Grenville's attraction to the independents vanished altogether. On the contrary, John Brooke's analysis of the divisions in 1769 has shown that almost 50 per cent of those elected to the House for the first time the previous year cast their votes in opposition on the Middlesex election.[78] Many of the defections that Grenville suffered in these debates were not immediately felt because of declarations of support from new MPs like Constantine Phipps (MP for Lincoln) and Chase Price (MP for Radnor). Yet this did not mean Grenville's troubles were over. The subtle difference between independent support in the old and new parliament was that he no longer knew how this group would react to his campaign on the Middlesex election or to which opposition party they owed their allegiance. The loyalties that Grenville had built up in office and his early years of opposition had simply lost their meaning by 1769. From this point on his campaign would have to be built on the thirty or so MPs whose allegiance was not in doubt.

[78] *The Chatham Administration*, p. 352.

To his credit, Grenville was not slow to recognize the changing nature of his support. The one constant factor in the divisions of 2 and 3 February had been the solid support of his closest followers, and it was on their backs that the campaigns of the next twelve months were borne. Grenville made two crucial decisions before the end of the session to ensure that his party took a leading part in the dispute over the Middlesex election. First, he reopened direct negotiations with the Rockinghams. They began in early February and Grenville himself handled the consultations with Dowdeswell, their leader in the Commons. To the obvious delight of both parties, common ground was immediately established and ideas for co-operation on a wider front were soon being aired.[79] Second, he took a much more active rôle in the day-to-day running of his party. This surprised many of his friends, but it was based solely on the issue at hand. There was no doubt in Grenville's mind that the decision to deprive Wilkes of his seat and return Luttrell in his place was a constitutional error of the highest order, and should be opposed at every opportunity. He spared no effort in preparing speeches to this end and whipping in MPs for the two crucial divisions. As he told Clive shortly before the debate on 8 May, 'there will be great business brought on in the House of Commons at the beginning of next week, where Mr. Grenville thinks Lord Clive's attendance as well as his own will be necessary'.[80]

It was a timely intervention on both counts. Co-operation with the Rockinghams at Westminster undoubtedly helped to fan the strong feeling of injustice at Luttrell's election in Wilkes's stead. On 3 February the minority was a respectable 137 votes, and this soon improved as government policy on the expulsion moved into its final phase. In the division on 15 April declaring Luttrell the legally elected member for Middlesex the minority increased to 143, and on 8 May reached the creditable total of 152. There was no chance of bringing the government down in these debates but the opposition did establish a degree of unity that survived intact until Grenville's death. The day after Luttrell's election

[79] *Grenville Papers*, iv. 411–12.
[80] Powis MSS G 37, Box 57.

was confirmed, Rockingham and Grenville dined together at the Thatched House Tavern and decided to take the fight against the government's action into the country. It was the first time that Grenville had ever contemplated political action outside Westminster and bears witness to the heat generated in the debates of that session.[81] Any doubts he may have felt at the decision were soon forgotten as the electors responded enthusiastically to the petitions against Luttrell's election. That Grenville could commit his followers to this adventurous policy was also due in no small part to the firm grip he had exerted over the party since early February. Votes recorded in the debates of February, April, and May indicate clearly that each member had been made aware of the need to register his opposition in this 'great business'. On 2 February Grenville had made a good start to the campaign, carrying two-thirds of his party into the opposition lobby, and this standard was not allowed to slip. On 3 February the Grenvillite vote totalled twenty-four; on 15 April this increased to twenty-six and in the last division of 8 May was raised again to thirty-one.

The party rejoiced at these figures. It had not only maintained its share of the minority but shown a marked improvement at the close of the session. True, there was something hollow about the celebrations. The numbers represented no more than 16 per cent of the minority vote in contrast to the 25 per cent achieved in the old parliament. Yet in the wake of the upheavals since the general election, Grenville was grateful for proof that his party was at least holding its own. Of particular comfort was the knowledge that this position had been attained in spite of some erratic voting behaviour by several of his followers. Only fourteen Grenvillites had voted in all three divisions on Wilkes since 2 February: John Carnac, Clive, George Clive, George Grenville, Henry Grenville, William Gerard Hamilton, Thomas Hampden, William Hussey, Sir John Irwin, Thomas Pitt, Lord George Sackville, Henry Seymour, Henry Strachey, and Thomas Whately. Six had attended and voted in two of the divisions: Henry Crabb Boulton, Thomas Durrant, Thomas

[81] For further detail and information on this see Lawson, Ph.D. thesis, pp. 127–95.

Fenwick, Fife, Thomas Howard, and Thomond. Seven had entered the opposition lobby on only one occasion: William Clive, Richard Clive, Sir Robert Fletcher, James Harris, George Hobart, John Walsh, and Alexander Wedderburn. Two, Arthur Chichester and Thomas Orby Hunter, had taken no part in these divisions at all. It is difficult to explain this pattern. The examination of these divisions by Professor P. D. G. Thomas has shown that this casual attitude to attendance was common to government and opposition parties alike. In terms of numbers lost, however, it would seem that the opposition suffered most. Two hundred and two MPs voted in the minority in the three divisions but only ninety-two, roughly 40 per cent, were actually present at all three.[82] In Grenville's case there were more obvious reasons for the absence of one or two members. Thomas Orby Hunter, for example, was fatally ill and Wedderburn was simply awaiting the right moment to break with his patron Sir Laurence Dundas, now a supporter of the government, but the failings of the remainder can only be put down to apathy. If Grenville and Rockingham could have conquered this enemy on 15 April and 8 May the results of the divisions could have been quite different.

The seven months before parliament reassembled on 9 January 1770 were to be the most fruitful of Grenville's opposition career. Like the East India inquiry before it, the campaign against Luttrell's adoption drew Grenville and Rockingham into a much closer working relationship. In this instance, however, it survived the whole of the parliamentary recess. The vital impetus was provided by the petitioning movement that developed in response to the government's decision to seat Luttrell. The idea did not originate at Westminster. The Middlesex electors had presented a petition against Luttrell's return on 29 April which was later considered and rejected.[83] Opposition politicians merely supported the movement when it became expedient to exert some control over the context of the petitions. When Grenville and Dowdeswell met at the Thatched House Tavern on 9 May, their only fear was that pleas for

[82] *The Parliamentary Lists of the Early Eighteenth Century*, pp. 51–2.
[83] This was on 8 May when the Commons confirmed Luttrell's election.

parliamentary reform would eventually dominate the protest and ruin the chances of a successful opposition campaign. The movement provided a unique opportunity to maintain pressure on the government during the recess, and they had no intention of surrendering the initiative to the radicals. From the very beginning they decided to eliminate all the peripheral issues and restrict the protest to the illegality of Luttrell's adoption. It was a gamble but they believed that this above all else, should hold the widest appeal to both the electors and MPs. In view of the disparate groups involved in organizing the petitions, they had set themselves a difficult task. It entailed a diligent watch on the county meetings and random gatherings that launched the petitions, and, more important, a frequent exchange of information between the two parties. Nevertheless they proved equal to the challenge. In certain petitions, requests for a dissolution did appear but the deviation went no further than this.

Grenville himself revelled in the job of co-ordinating his party's efforts during these months. Though his movements were restricted due to his wife's serious illness, he played a crucial role in the preparation of several petitions, advising on tactics and answering legal queries concerning their content. He also maintained a firm grip on the party's strategy for the coming session. At a vital meeting of the inner circle in October it was decided to use the petitioning movement as a launching-pad for a full-scale attack on the ministry. There were reservations about this course of action. Grenville in particular was not happy with the idea of employing an opposition movement outside parliament to bring pressure to bear on the administration. In the end, he told Whately, it was the dire state of the nation's fortunes that had really forced his hand. As an opposition MP he may have been 'perfectly innocent' of the troubles around him, but it was his duty, through his best endeavours, 'to prevent this ruin'.[84] True to his word, Grenville made every effort to prepare his followers for the new session. The whips were sent out, and plans for the opening debates of the session carefully concerted with the Rockinghams.

[84] Grenville's Letterbook, vol. 2 (26 Nov. 1769).

Expectations of early success did not materialize. The death of Grenville's wife in December completely disrupted preparations for the opening debates, and temporarily left the party leaderless. Grenville promised his followers that he would do all in his power to attend on 9 January, but he was in no condition beforehand to finalize details for opposition to the Address.[85] This was certainly evident in the attendance at the opposition's eve of session dinner in the Thatched House Tavern. Of the eighty-one present, the Grenvillites could muster only fourteen MPs.[86] Not surprisingly, the situation improved drastically the following day when Grenville reached London. Though he only presented a low profile in the debate on the Address, his party's share of the minority vote was much healthier than presence at the dinner had indicated. At this stage Grenville's voting strength was twenty-seven, and eighteen of these were in the minority on Dowdeswell's amendment for an inquiry into the discontents throughout the kingdom.[87] Fifteen had voted in opposition on 8 May 1769: George Clive, Sir Robert Fletcher, George Grenville, Henry Grenville, Thomas Hampden, William Gerard Hamilton, Thomas Howard, William Hussey, Sir John Irwin, Thomas Pitt, Lord George Sackville, Henry Seymour, Henry Strachey, Thomond, and Thomas Whately. Two, George Hobart and John Walsh, had registered only one opposition vote since 3 February 1769; and one, Arthur Chichester, had been absent from the House throughout all the proceedings on the Middlesex election. In addition six of Grenville's independent supporters from the old parliament were numbered in the minority: Sir Brook Bridges, Thomas Grosvenor, John Rushout, Robert Waller, John Rolle Walter, and Thomas Willoughby. Overall it was a disappointing debate for the opposition. The government won the division by the extremely comfortable margin of 116 votes.[88] Yet there was just one crumb of comfort for Grenville. Despite his incapacity, the party's share of the minority vote had

[85] *Chatham Correspondence*, iii. 390.

[86] The full list can be found in Rockingham's papers W.W.M. R1-1259. It is misdated by Namier and Brooke (*House of Commons*, i. 528) as 8 June 1770.

[87] Almon, *Debates*, viii. 175-7.

[88] *Commons Journals*, xxxii. 456.

not fallen below the 16 per cent achieved the previous spring.

It seems likely that Grenville noted this fact because the division marked the end of his mourning. The following week he fully resumed his position as leader of the party. Two factors had made this move imperative. First, the opposition intended to continue the campaign on the Middlesex election by moving for a Committee on the State of the Nation on 23 January. Second, despite its success on the opening day of the session, the government was wracked with internal divisions. The departure of Camden, the lord chancellor, and Granby, master of the ordnance, from the cabinet had, as Walpole observed at the time, left affairs in 'a critical situation'. To add to his troubles, Grafton was experiencing great difficulty in finding suitable replacements. Charles Yorke, his first choice as chancellor, simply could not decide whether to take office or remain in opposition. It was an unenviable dilemma, and if it continued unresolved Walpole could see only two avenues of escape; 'a change of administration and a dissolution of Parliament'.[89] Grenville and Rockingham would have welcomed either of these alternatives, and they pressed hard during the next three weeks to force the government into submission. The first serious test of nerve came in the Committee on the State of the Nation on 25 January when the opposition sought to rescind all previous resolves on Luttrell's election at Middlesex. After a long and heated debate in a very full House the opposition ran the government to within forty-four votes of defeat.[90]

In his analysis of the debate Walpole put the opposition advance down to Camden's dismissal and 'much maladroitness in stating the question on the court side'.[91] This in itself cannot be denied, but the interpretation underestimates the detailed preparation and planning undertaken by the opposition. Not only had they managed to whip the vast majority of their supporters into the House but, more significant, struck on a legal point that MPs found it difficult to refute. Grenville was delighted with the response of his own followers. All those who had voted in the minority on the

[89] *Letters*, vii. 352–4.
[90] The figures were 224 to 180 votes.
[91] *Letters*, vii. 361.

Address were again present with the addition of Clive, Fife, and Wedderburn.[92] The independents were similarly enthusiastic. Of the nine supporters Grenville had carried into opposition from the old parliament only Sir Brook Bridges did not appear in the minority. It is unfortunate that this is the last-known division list before Grenville's death. The opposition campaign in the Committee on the State of the Nation continued until 19 February but no indication of Grenvillite attendance or voting record has been preserved. This is certainly a matter for regret. On 31 January the opposition reduced the government's majority to forty votes and it can only be assumed that MPs in the minority were unchanged from the week before.[93] The lack of information is of no less importance when the debates of February are scrutinized. In two divisions during this month North, who had succeeded Grafton as first lord, frustrated attempts to defeat the government on the floor of the House, but it is impossible to construct a full picture of the drama. The division figures remain, and show that slowly but surely North increased the government's majority. Yet no record of those returning to the government fold survives.

To make an assessment of Grenville's political strength during the remaining weeks of the session, therefore, is largely a matter of guesswork. The solitary contemporary source of any value does indicate that it remained intact and unaffected by the government's resurgence under North. On 22 March 1770 Lord Mayor Beckford held an opposition dinner at the Mansion House and a list of those in attendance was published in an edition of *Lloyd's Evening Post*.[94] The most notable feature of the gathering was the presence of opposition politicians from both Houses. Six of Grenville's leading supporters in the Lords attended the dinner: Coventry, Fortescue, Hyde, Lyttelton, Suffolk, and Temple.[95] This represented half his strength in the Lords, and they were accompanied by seventeen Grenvillites from the Commons.

[92] The minority list appears in Almon, *Debates*, viii, 190–2.

[93] The figures on this day were 226 to 186 votes.

[94] The first list appeared in the issue covering 21 to 23 March 1770, and a second appeared in the issue for 22–26 March 1770.

[95] There is a list of Grenville's supporters in the Lords, apparently prepared by Newcastle, in Rockingham's papers W.W.M. R1–786. Their activities in this period are discussed by Lowe, Ph.D. thesis, pp. 493–528.

Grenville himself was absent but sent his son, Thomas, who was not yet an MP, to represent him. The whole group apart from Grenville's son had voted in the minority of 25 January. There seems no reason to contradict the evidence of continuity suggested by this list. The remaining weeks of the session were an anticlimax after the divisions of January. Attempts to open campaigns on America and excessive spending against the civil list did not prove successful. Under North's leadership the government maintained a firm command of business in the Commons, and on the very last day of the session he crushed a well-planned attack against the ministry's colonial policies by the margin of 199 to 79 votes.[96]

Yet Grenville did achieve one spectacular success in these weeks with the passage of his bill to reform the trials of controverted elections. Moreover, it was done without co-operation with either the Rockinghams or his own followers, and in the face of ministerial opposition. The abuses surrounding the hearing of petitions contesting election returns were well known to Grenville s contemporaries. Since 1672 petitions had been heard before the whole House and the chances of receiving an impartial judgement were minimal. The government of the day, for example, could use its influence to reach a quick decision in its favour or, at the other extreme, delay proceedings so long that the petitioner simply ran out of time and money, finally withdrawing the case. The worst aspect of all surrounding these hearings was the tedium. Debates on contested election returns were always dominated by lawyers, and persuading MPs to attend sittings and listen to speeches few of them understood proved extremely difficult. Grenville knew of these abuses, but his motives for assuming the responsibility to reform them are obscure. He had never himself been involved in a contested election return, and only one or two of his friends had suffered this ordeal. The real catalyst appears to have been the Wilkes case of 1769. Here was an instance of the Commons sitting in judgement on who should be elected to the House, and rejecting the lawfully returned candidate. Worse still,

[96] This division took place on 9 May and concerned eight resolutions censuring the conduct of the administration with regard to America, *Commons Journals*, xxxii, 970.

the government allowed Henry Lawes Luttrell to be seated in his place for Middlesex, even though the electors had rejected him. Of all the abuses in election cases of the 1760s this was the most flagrant example of a biased verdict. Thus when Grenville introduced his bill to the Commons on 28 February 1770 it received a warm reception. As Walpole observed afterwards, it was a measure 'the profligacy of the times loudly demanded'.[97]

For what was in effect a private member's bill, the legislation passed through parliament remarkably quickly, becoming law in April 1770. The reasons for this were twofold. First, Grenville insisted throughout its passage that the bill be treated as a non-party measure. His standing and prestige in the House was such that MPs accepted his explanation that his reforms were motivated purely by a desire to restore the House's 'dignity and honour' in election cases.[98] Second, the proposals themselves offered MPs an honourable escape from the drudgery of attending election hearings. In its final form Grenville's Election Act established a Select Committee of the House, drawn at random, to hear petitions on contested returns. This removed the process from the Commons chamber and freed valuable debating time for weightier matters of state.[99] The reaction of MPs was so enthusiastic that it overwhelmed the government's efforts at baulking the bill's passage. North saw in Grenville's proposals an attempt to resurrect the Wilkes issue and hoped to knock the bill on its head at birth. But the favourable response of the independents, and even members of the government, persuaded North to give way on 2 April and allow the proposed reforms to pass. The first few years of the bill's existence were intended as a trial period. However, the immediate success of the legislation in devising a means of fairer hearings and judgements guaranteed its survival well into the next century. The act was a fitting tribute to Grenville's unstinting duty at Westminster throughout his life.

The success of guiding this measure on to the statute-book to the general opposition campaign is more difficult

[97] *Memoirs of George III*, iv. 74.

[98] Almon, *Debates*, viii. 240.

[99] For a full account of the Act and its passage see Lawson, 'Grenville's Election Act 1770', *B.I.H.R.* 128 (1980), 218–28.

to ascertain. The failure to topple the ministry in January and February 1770 was to haunt opposition leaders for many years to come. Perhaps the best that could be said of the Election Act in this context is that it provided a much-needed boost to morale after North had dealt the opposition campaign a severe blow. The personal praise was certainly a welcome note for Grenville. When he retired for the summer recess his health was failing and he found it a hard task simply answering his correspondence. If his supporters were aware of this physical decline they kept a discreet silence. There were no complaints about his inactivity and everyone remained confident that Grenville would speak in the opening day of the new session. That they remained true to the end is no surprise. It was this unswerving loyalty that Grenville inspired in the hard core of his following which lies at the root of his success as an opposition leader. He did not possess the wealth or aristocratic influence to bind his followers to the opposition cause, and yet he found that personal devotion to the interests of each party member achieved the same result. The point was made most forcibly by attorney-general Edward Thurlow in 1774. He had just refused to carry out cabinet proposals to punish the culprits of the Boston Tea Party and made no secret of the motive. 'They want to throw the whole responsibility of the business upon the Solicitor-General and me', he told a subordinate, 'and who would be such damned fools as to risk themselves for such . . . fellows as these. Now if it was George Grenville, who was so damned obstinate that he would go to hell with you before he would desert you, there would be some sense in it.'[100] Allied to this sound character was a degree of political skill and acumen unrivalled by his opponents. He was a natural leader in the Commons, and his party prospered in a period when ministerial power resided in the Lords. Grenville certainly knew his strength lay in the debating chamber and pressed home the advantage with a consistency that the Rockinghams found hard to match. Even Walpole's unflattering epitaph could not conceal the reasons behind the prominence of Grenville and his followers during these years. 'Mr. Grenville was', he wrote, 'confessedly,

[100] H.M.C. Knox, p. 270.

the ablest man of business in the House of Commons, and, though not popular, of great authority there from his spirit, knowledge, and gravity of character.'[101] The concept of personal party was epitomized by Grenville's career in the years 1765–70.

[101] *Memoirs of George III*, iv. 125.

Postscript

Grenville died on 13 November 1770, the opening day of the new session of parliament. He had been ill since the summer and in October was carried to London without hope of recovery.[1] His followers knew of this physical decline but few voiced their reservations about the future at Westminster without him. Grenville was kept informed of the growing storm over the Falkland Islands but no plans were laid for the debate on the Address. The air was one of pervading gloom. 'I am sincerely concerned at the bad account you send me of Mr. Grenville's health', Lord George Sackville, now Germain, told Sir John Irwin on 23 October, 'his natural or political death would be the greatest misfortune to his friends and a real loss to this country.'[2]

Were these fears about the future well founded? Certainly no one had been designated to lead the party at Westminster in the event of Grenville's death. In the two studies where the fate of the Grenvillites after 1770 merits discussion two candidates are proffered, and both were in the Lords. The first, by D. A. Winstanley, is the Earl of Suffolk; and the second, by John Brooke, is Earl Temple.[3] Of the former there is no evidence in the correspondence of Grenville or his friend to support the assertion that Suffolk 'was commonly looked upon as Grenville's successor'.[4] Suffolk was a keen admirer of Grenville but not of opposition politics. Since the campaign on Wilkes began in the spring of 1769 he had taken no active part in opposition on this vital issue. Grenville exchanged pleasantries in correspondence with Suffolk until his death but they never agreed on a common approach to the question of Wilkes's expulsion and the Middlesex

[1] Walpole gives a graphic description of Grenville's ailments in *Memoirs of George III*, iv. 125.

[2] H.M.C. Stopford-Sackville MSS i. 131.

[3] *Lord Chatham and the Whig Opposition*, pp. 404–7 and *The Chatham Administration*, p. 29.

[4] *Lord Chatham and the Whig Opposition*, p. 405.

election.[5] Temple's candidature is raised and then dismissed by Brooke himself. He had the credentials for leadership: wealth, influence, and devotion to the opposition cause at Westminster. Yet his vanity and pomposity repelled would-be adherents. He seemed to revel in opposition for its own sake and it would have been folly to expect MPs to follow a man who so readily scorned office.

Potential leaders amongst Grenville's followers in the lower House are harder to identify. Thomas Whately, Grenville's 'man of business' had most knowledge of policies and personalities within the party but least talent. He was a dull methodical man with no flair at all as speaker in the Chamber. Of the four ablest debaters, Lord George Germain, Thomas Pitt, Henry Seymour, and Alexander Wedderburn, only Germain could be considered as a serious candidate for the leadership. He was rich and influential with an abrasive manner in the House that suited opposition politics. What he lacked was a circle of admirers. Apart from close ties with Sir John Irwin, and to a lesser extent, Wedderburn, Germain never established a base in the Commons on which to stake his claim as Grenville's successor. Wedderburn, on the other hand, had many admirers. He was attentive to the business of the House, and had a reputation as one of the best speakers in the Commons. His failings were poverty and an obsessive desire to become lord chancellor. Wedderburn never contemplated the task of leading the party in the Commons. Grenville's death released him from his allegiance to the party, and his instincts were soon directing him towards the government where his ambitions could be fulfilled. The careers of Pitt and Seymour never reached such heights. Both were good speakers and remained close to Grenville throughout his years in opposition, but neither had the necessary credentials for the post of leader. Pitt's political ambitions were hindered by domestic problems. In 1768 his financial difficulties were such that he was forced to sell one seat under his control in Cornwall to John Crauford, a government supporter. Seymour's position

[5] See, for example, Suffolk's letter to Grenville of 19 August 1769, *Grenville Papers*, iv. 437. There had been previous disagreements too. On 13 July 1766 Suffolk pleaded with Grenville then not to join the general opposition campaign against the new ministry; S.T.G. Box 22 (53).

was even weaker. He relied for his return at Huntingdon on his half-brother, Sandwich, who was then in government. Had his pretensions to leading an opposition party been realized, he could doubtless have been removed from the Commons altogether.

One candidate who did not press his claims but could almost certainly have assumed Grenville's mantle was Clive. There was no question-mark over his ability as an inspirational leader. He already possessed a small and loyal following at Westminster and had the wealth to extend his electoral interest for the benefit of the Grenville party. The flaw in Clive's candidature was his own lack of ambition. His following in parliament was established to mark his rise up the social ladder. Clive enjoyed the trappings of power, country estates, and electoral influence, but had no desire to govern. Indeed, his volatile character was ill-suited to the rigours of administering a ministerial office. Wedderburn informed Clive of Grenville's death on 14 November and made discreet enquiries then of his future intentions. 'Whatever part I may take in his conjuncture', he wrote, 'will never be decided without the fullest communication with you.'[6] The answer was blunt and to the point. Clive forsook all claims to the leadership: 'Your delicacy towards me serves only to convince me of the propriety of my conduct in leaving you the absolute master of your own conduct in Parliament, free from all control but that of your own judgement.'[7]

The Grenvilles were thus a leaderless flock when they assembled in London for the opening of the new session on 13 November 1770. Their last collective decision was to absent themselves from parliament on the Address as a mark of respect to their former leader. From that moment on the party was adrift. Germain and Wedderburn did attempt to rally Grenville's followers to the opposition in late November. On 22 and 27 November both lent strong support to abortive motions by Dowdeswell seeking intelligence on the negotiations with Spain.[8] But their efforts went unrewarded. They were not attracting previous adherents

[6] Cited in Malcolm, *Life of Lord Clive*, ii. 230.

[7] Clive to Wedderburn, 18 November 1770, ibid., p. 232.

[8] Accounts of these debates can be found in Eg. 222, pp. c.100–234 and Eg. 223, pp. 1–45 respectively.

to Grenville's cause, and in the first division list after his death the minority boasts only five party members: John Carnac, William Gerard Hamilton, Sir John Irwin, Lord George Germain, and Henry Seymour.[9] The end came in the new year after the Duke of Bedford's death on 14 January. It gave North the opportunity to strengthen the ministry and he wasted no time in approaching the Grenvilles. The intermediary was Suffolk. On 22 January he accepted the office of privy seal and sought preferment for his colleagues. North, at that time, could find places for only two of Grenville's followers in the Commons: Whately became a lord of trade and Wedderburn was appointed solicitor-general. The prospect of office, however, was enough to persuade four others to transfer their allegiance to government straight away: Lord Fife, Sir Robert Fletcher, James Harris, and George Hobart.

This was not a mass defection. When the House debated the Address of Thanks for a convention with Spain on 13 January 1771 the Grenville party on a full muster still had over twenty members. Its ranks were depleted but with Clive, Germain, Seymour, and Pitt remaining in opposition the fate of the party was not yet sealed. Indeed, in the vote on 13 February, Grenville's supporters, including independents like Thomas Grosvenor and Fane William Sharpe, numered seventeen.[10] It was an encouraging trend and the rump of Grenville's following continued this good showing until the Easter recess. The impetus was provided by the speeches of Germain and Seymour, who not only launched attacks on ministerial policy but, more important, harassed and castigated the defectors. 'I knew what Grenville thought of the present Administration', Seymour told them on 27 March, 'I never could have joined them without being able to convince myself that was he alive he would do so too.'[11] On a matter of principle such attacks were well founded; in the spring of 1768 Grenville had sworn, with

[9] *Middlesex Journal*, 24–26 January 1771. This was on a motion for a committee to inquire into the administration of criminal justice in Westminster Hall. The voting figures were 76 to 184.

[10] The full list of the minority is given in Almon, *Debates*, ix. 142–4.

[11] Cited in Namier and Brooke, *History of Parliament*, iii. 423. For a more personalized attack on Wedderburn see Walpole, *Memoirs of George III*, iv. 203.

Whately's knowledge, never to serve in government with a Bedford.[12]

Yet these tactics enjoyed only a limited success. They were dependent on active opposition at Westminster and after parliament rose in May alliances within the party crumbled away. In part this was a natural reaction to Grenville's death. To MPs personally connected with Grenville, like Thomond and Arthur Chichester, opposition held little appeal after November 1770. They drifted apart from their colleagues and rarely attended the House during the remainder of their parliamentary careers. More important, however, was the loss of Whately's organizational talents. His defection rather than Wedderburn's must have given North most satisfaction. He was the only follower of Grenville who could have reconciled conflicting interests within the party. It was Whately's work in the months of inactivity at Westminster that had bound Grenville's adherents to the opposition over the previous five years. Once he departed so did the will to continue as a unit. Clive and his group joined the ministry in September 1772 thus putting the final nail in the coffin. Of those not lost at the 1774 general election only Lord George Germain, William Gerard Hamilton, Thomas Howard, William Hussey, Sir John Irwin, Thomas Pitt, and Henry Seymour were still in opposition in the new parliament. They were no longer seen as Grenvillites; this species had perished with its progenitor four years earlier.

[12] *Grenville Papers*, iv. 274.

Bibliography

A. Manuscripts

i. *In the British Museum*

Egerton MSS.	215-263, 3711
Add. MSS.	33002, 33037, 32980, 32981, 32987
Add. MSS.	42083-42088
Add. MSS.	57804, 57807, 57810, 57811, 57817A, 57818, 57820

ii. *In other repositories*

Buccleuch	MSS.	Scottish Record Office
Burke	MSS.	Sheffield City Library
Bute	MSS.	Cardiff City Library
Clive	MSS.	National Library of Wales
Dunster Castle	MSS.	Somerset Record Office
Glynllifon	MSS.	National Library of Wales
Guildhall	MSS.	Sandwich-Webb Correspondence, Guildhall Library, London
Minto	MSS.	National Library of Scotland
Newdegate	MSS.	Warwickshire County Record Office
Portland	MSS.	Nottingham University Library
Powis	MSS.	India Office Library
Stowe	MSS.	Henry E. Huntingdon Library, ST 7-Grenville Letterbooks vols. 1 and 2; STG – General Correspondence.
Strutt	MSS.	Essex County Record Office
Upton	MSS.	Leeds City Archive
Wentworth Woodhouse Muniments		Sheffield City Library
Willis	MSS.	Bodleian Library
Zetland	MSS.	North Yorkshire County Record Office

iii. *In private possession*

Oswald	MSS.	Hockworthy House, Somerset
Ryder	MSS.	Sandon Hall, Stafford

B. Printed Sources

i *Official and Parliamentary Sources*

Journals of the House of Commons (cited as *Commons Journals*)
Journals of the House of Lords (cited as *Lords Journals*)

Almon, J., *The Debates and Proceedings of the British House of Commons from 1743 to 1774* (11 vols., 1766–75; cited as Almon, *Debates*).

Cobbett, W., *Parliamentary History of England from . . . 1066 to . . . 1803* (36 vols., 1806–20; cited as *Parliamentary History*).

ii. *Contemporary Correspondence and Memoirs*

Historical Manuscripts Commission (cited as H.M.C.)
8th Report, Digby MSS.
10th Report, I, Weston-Underwood MSS.
12th Report, X. Charlemont MSS.
Carlisle MSS.
Egmont MSS. (cited as *Egmont Diary*)
Foljambe MSS.
Lothian MSS.
Stopford-Sackville MSS.
Various MSS. VI (Knox).

Correspondence of John, Fourth Duke of Bedford, selected from the originals at Woburn Abbey, ed. Lord J. Russell (3 vols., London, 1842–6; cited as *Bedford Correspondence*).

The Correspondence of Edmund Burke, ed. T.W. Copeland (10 vols., Cambridge, 1958–78; cited as *Burke Correspondence*).

The Works of the Right Honourable Edmund Burke (12 vols., London, 1849).

Letters from George III to Lord Bute 1756–66, ed. R. Sedgwick (London, 1939).

Selections from the Family Papers preserved at Caldwell, ed. W. Mure (2 vols., Glasgow, 1854; cited as *Caldwell Papers*).

Documents Relating to the Constitutional History of Canada, 1759–91, ed. A. Shortt and A.G. Doughty (2nd edition, Ottawa, 1918).

Correspondence of William Pitt, Earl of Chatham, ed. W.S. Taylor and J.H. Pringle (4 vols., London, 1838–40; cited as *Chatham Correspondence*).

The Life of Robert, Lord Clive: Collected from the Family Papers Communicated by the Earl of Powis ed. Sir J. Malcolm (3 vols., 1836).

The Political Journal of George Bubb Dodington, ed. J. Carswell and L.A. Dralle (Oxford, 1965; cited as *Dodington Diary*).

Henry Fox, First Lord Holland His Family and Relations, ed. Earl of Ilchester (2 vols., London, 1920).

The Papers of Benjamin Franklin, ed. L.W. Labaree (New Haven and London, 1959–   ; cited as *Franklin Papers*).

The Fitch Papers. Correspondence and Documents During Thomas Fitch's Governorship of the Colony of Connecticut 1754–66. (Connecticut Historical Society Collections, 18, 1920).

The Private Correspondence of David Garrick with the most celebrated persons of his times . . ., ed. J. Boaden (2 vols., London, 1831; cited as *Garrick Correspondence*).

The Correspondence of King George the Third from 1760 to December 1783, ed. Sir. J. Fortescue (6 vols., London, 1927–8; cited as *Fortescue*).

Richard Glover: Memoirs by a Celebrated Literary and Political Character from . . . 1742 to . . . 1757 (London, 1814; cited as Glover, *Memoirs*).

Autobiography and Political Correspondence of Augustus Henry, Third Duke of Grafton K.G., ed. Sir. W.R. Anson (London, 1898; cited as *Grafton Autobiography*).

The Grenville Papers; being the Correspondence of Richard Grenville Earl Temple K.G., and the Right Hon. George Grenville their Friends and Contemporaries, ed. W.J. Smith (4 vols., London, 1852–3).

Additional Grenville Papers, 1763–65, ed. J.R.G. Tomlinson (Manchester, 1962).

'A Selection from the Correspondence and Miscellaneous Papers of Jared Ingersoll', ed. F.B. Dexter (*Papers of New Haven Colony Historical Society*, 9, 1918; cited as *Ingersoll Papers*).

The Jenkinson Papers, 1760–66, ed. N.S. Jucker (London, 1949).

The Letters of Junius, ed. J. Cannon (Oxford, 1978).

'Leicester House Politics, 1750–60, from the papers of John, second earl of Egmont', ed. A.N. Newman (*Camden Miscellany*, xxiii, fourth series, vii).

Memoirs and Correspondence of George, Lord Lyttelton from 1734 to 1773, ed. R.J. Phillimore (2 vols., London, 1845; cited as *Lyttelton Memoirs*).

A series of Letters of the First earl of Malmesbury his family and friends from 1745 to 1820, ed. The earl of Malmesbury (2 vols., London, 1870).

A Selection from the Papers of the Earls of Marchmont in the possession of the Right Hon. Sir George Henry Rose . . . 1685–1750 (3 vols., London, 1831; cited as *Marchmont Papers*).

The Border Elliots and the Family of Minto, ed. G.F.S. Elliot (Edinburgh, 1897).

Memorials of the Public Life and Character of James Oswald of

Dunnikier . . ., ed., Rt. Hon. J. Oswald (Edinburgh, 1825; cited as Oswald, *Memorials*).

Memoirs of the Administration of the Rt. Hon. Henry Pelham, collected from the family papers, and other authentic documents (2 vols., London, 1829; cited as Coxe, *Pelham*).

Memoirs of the Marquis of Rockingham and His Contemporaries, ed. George Thomas, Earl of Albemarle (2 vols., London, 1852; cited as *Albemarle*).

'The Parliamentary Diaries of Nathaniel Ryder 1764–67', ed. P.D.G. Thomas (*Camden Miscellany*, xxiii, fourth series, vii; cited as *Ryder Diary*).

The Fourth Earl of Sandwich: Diplomatic Correspondence 1763–65, ed. F. Spencer (Manchester, 1961).

George Selwyn and his contemporaries; with memoirs and notes; ed. T.H. Jesse (4 vols., London, 1843–4).

Life of William Earl of Shelburne, afterwards First Marquis of Lansdowne with Extracts from his Papers and Correspondence, ed. Lord Edmond Fitzmaurice (3 vols., London, 1875–6; cited as *Shelburne*).

James, Earl Waldegrave, Memoirs from 1754 to 1758 (London, 1821; cited as *Waldegrave Memoirs*).

The Letters of Horace Walpole, Fourth Earl of Orford, ed. Mrs. Paget Toynbee (16 vols., London, 1903–5; cited as Walpole, *Letters*).

Horace Walpole's Memoirs of the Reign of George II, ed. Lord Holland (3 vols., 1846).

Horace Walpole, Memoirs of the Reign of King George the Third, ed. G.F. Russell Barker (4 vols., 1894).

Memoirs of the life and administration of Sir Robert Walpole, earl of Orford (3 vols., London 1798; cited as Coxe, *Walpole*).

Chronicles of the Eighteenth Century Founded on the Correspondence of Sir Thomas Lyttelton and his family, ed. M. Wyndham (2 vols., London, 1924).

The Life and Correspondence of Philip Yorke, Earl of Hardwicke, Lord High Chancellor of Great Britain, ed. P. Yorke (3 vols., Cambridge, 1913; cited as Yorke, *Hardwicke*).

iii. *Periodicals and Pamphlets*

(a) *Periodicals*

Gazetteer
Gentleman's Magazine
London Chronicle
Middlesex Journal
North Briton
Political Register

(b) *Pamphlets*

The Controversy Between Great Britain and the Colonies Reviewed (London, 1769).

The Present State of the Nation (London, 1768).

The Regulations Lately Made Concerning the Colonies and the Taxes imposed upon them, considered (London, 1765).

Thoughts on the Present Discontents (London, 1770).

Pollard, A. F., *Political Pamphlets* (London, 1897).

SECONDARY WORKS

A. Books

Almon, J., *Anecdotes of the Life of the Rt. Hon. William Pitt, Earl of Chatham* (6th ed., 3 vols., London, 1797).

Barrow, Sir J., *The Life of George, Lord Anson* (London, 1839).

Barrow, T. C., *Trade and Empire: The British Customs Service in Colonial America 1660–1775* (Camb., Mass., 1967).

Baugh, D., *Naval Administration in the Age of Walpole* (Princeton, N. J., 1965).

Bonsall, B., *Sir James Lowther and Cumberland and Westmorland Elections 1754–1775* (Manchester, 1960).

Brewer, J., *Party Ideology and Popular Politics at the Accession of George III* (Cambridge, 1976).

Brewer, J. and Styles, J., eds., *An Ungovernable People* (New Brunswick, N.J., 1980).

Brooke, J., *King George III* (London, 1972).

Brooke, J., *The Chatham Administration 1766–1768* (London, 1956).

Browining, R., *The Duke of Newcastle* (London, 1975).

Buckinghamshire, *Victoria County History* (4 vols., 1905).

Christie, I. R., and Labaree B. J., *Empire or Independence, 1760–1776* (Oxford, 1976).

Clark, J. C. D., *The Dynamics of Change* (Cambridge, 1982).

Colley, L., *In Defiance of Oligarchy: The Tory Party 1714–60* (Cambridge, 1982).

Ernst, J. A., *Money and Politics in America, 1755–1775; a study in the Currency Act of 1764 and the political economy of revolution* (Chapel Hill, N.C., 1973).

Foord, A. S., *His Majesty's Opposition, 1714–1830* (Oxford, 1964).

Gipson, L. H., *The British Empire before the American Revolution, vol. X* (New York, 1961).

Langford, P., *The Excise Crisis* (Oxford, 1975).

Langford, P., *The First Rockingham Administration 1765-1766* (Oxford, 1973).

McKelvey, J. L., *George III and Lord Bute; The Leicester House Years* (Durham, N.C., 1973).

Namier, Sir L. B. and Brooke, J., *Charles Townshend* (London, 1964).

Namier, Sir L., B. and Brooke, J., eds, *The House of Commons 1754-1790* (3 vols., London, 1964).

Newman, A., eds., *The Parliamentary Lists of the early eighteenth century: their compilation and use* (Leicester, 1973).

Nobbe, G., *The North Briton. A study in Political Propaganda* (New York, 1939).

O'Gorman, F., *The Rise of Party in England: The Rockingham Whigs 1760-1782* (London, 1975).

Owen, J. B., *The Eighteenth Century 1714-1815* (London, 1975).

Owen, J. B., *The Rise of Pelhams* (London, 1957).

Peters, M., *Pitt and Popularity* (Oxford, 1980).

Perry, T. W., *Public Opinion, Propaganda and Politics in Eighteenth Century England* (Camb., Mass., 1962).

Plumb, Sir J. H., *Sir Robert Walpole* (2 vols., London, 1956-60).

Rashed, Z. E., *The Peace of Paris, 1763* (Liverpool, 1951).

Rea, R. R., *The English Press in Politics, 1760-1774* (Lincoln, Neb., 1963).

Richmond, H. V., *The Navy in the War of 1739-48* (3 vols., Cambridge, 1920).

Sedgwick, R., ed., *The House of Commons 1715-1754* (2 vols., London, 1971).

Shy, J., *Toward Lexington: The Role of the British Army in the Coming of the American Revolution* (Princeton, 1965).

Sosin, J., *Whitehall and the Wilderness: The Middle West in British Colonial Policy 1760-1775* (Lincoln, Nebraska, 1961).

Speck, W. A., *Stability and Strife* (London, 1977).

Sutherland, Dame L. S., *The East India Company in Eighteenth Century Politics* (Oxford, 1952).

Thomas, P. D. G., *British Politics and the Stamp Act Crisis: The First Phase of the American Revolution 1763-1767* (Oxford, 1975).

Thomas, P. D. G., *Lord North* (London, 1976).

Thomas, P. D. G., *The House of Commons in the Eighteenth Century* (Oxford, 1971).

Thomson, M. A., *The Secretaries of State 1681–1782* (Oxford, 1932).

Ubbelohde, C., *The Vice-Admiralty Courts and the American Revolution* (Chapel Hill, 1960).

Walton, G. M., and Sheperd, J. F., *The Economic Rise of Early America* (Cambridge, 1979).

Western, J. R., *The English Militia in the eighteenth century: the story of a political issue, 1600–1802* (London, 1965).

Wiggin, L. M., *The Faction of Cousins: A Political Account of the Grenvilles, 1733–1763* (New Haven, 1958).

Wilkes, J. W., *A Whig in Power: The Political Career of Henry Pelham* (Chicago, 1964).

Winstanley, D. A., *Lord Chatham and the Whig Opposition* (Cambridge, 1912).

B. Periodicals

Brewer, J., 'The Misfortunes of Lord Bute: A Case Study in Eighteenth Century Political Argument and Public Opinion', *H[istorical] J[ournal]*, 16 (1973).

Browning, R., 'The Duke of Newcastle and the Imperial Election Plan, 1749–1754', *J[ournal of] B[ritish] S[tudies]*, II, 1 (1967).

Humphreys, R. A., 'Lord Shelburne and the Proclamation of 1763', *E[nglish] H[istorical] R[eview]*, 49 (1934).

Jarrett, D., 'The Regency Crisis of 1765', *English Historical Review*, 85 (1970).

Johnson, A. S., 'The Passage of the Sugar Act', *W[illiam and] M[ary] Q[uarterly]*, 3rd series, 16 (1959).

Langford, P., 'William Pitt and Public Opinion, 1757', *English Historical Review*, 88 (1973).

Lawson, P., 'Grenville's Election Act 1770', *B[ulletin of the] I[nstitute of] H[istorical] R[esearch]*, 128 (1980).

Lawson, P., 'George Grenville and America – The Years of Opposition 1765–1770', *William and Mary Quarterly*, 3rd series, 37 (1980).

Lawson, P., 'Parliament and the First East India Inquiry, 1967', *P[arliamentary] H[istory] Y[earbook] B[ook]*, 1 (1983).

Marshall, P., 'The Incorporation of Quebec in the British Empire,

1763-1774', *Of Mother Country and Plantations: Proceedings of the Twenty Seventh Conference in Early American History* eds., V. B. Platt and D. C. Skaggs (Bowling Green, Ke., 1971).

Middleton, R., 'Pitt, Anson and the Admiralty, 1756-1761', *History*, 55 (1970).

Morgan, E. S., 'The Postponement of the Stamp Act', *William and Mary Quarterly*, 3rd series, 7 (1950).

Namier, Sir L. B., 'Charles Garth and his Connections', *English Historical Review*, 54 (1939).

Namier, Sir L. B., 'The Circular Letters: An Eighteenth Century Whip to Members of Parliament', *English Historical Review*, 44, (1929).

Thomas, P. D. G., 'Charles Townshend and American Taxation in 1767', *English Historical Review*, 83 (1968).

Tracy, N., 'The Gunboat Diplomacy of the Government of George Grenville, 1764-1765: The Honduras, Turks Island and Gambian Incidents', *Historical Journal*, 17 (1974).

C. Unpublished University Theses

Clark, J. C. D., The Administration of the Duke of Newcastle 1754-56, (Ph.D. Cambridge, 1980).

Hardy, A., The Duke of Newcastle and his Friends in Opposition 1762-1765 (M.A. Manchester, 1956).

Lawson, P., Faction in Politics: George Grenville and his Followers 1765-1770 (Ph.D. Wales, 1980).

Lowe, W. C., Politics in the House of Lords, 1760-1775 (Ph.D. Emory Univ., Georgia, 1975).

Tomlinson, J. R. G., The Grenville Papers 1763-1765 (M.A. Manchester, 1956).

Index